# Smart Cities and Positive Energy Districts: Urban Perspectives in 2021

# Smart Cities and Positive Energy Districts: Urban Perspectives in 2021

Editor

**Paola Clerici Maestosi**

MDPI • Basel • Beijing • Wuhan • Barcelona • Belgrade • Manchester • Tokyo • Cluj • Tianjin

*Editor*
Paola Clerici Maestosi
ENEA—Italian National
Agency for New Technologies
Energy and Sustainable
Economic Development
Department Energy
Technologies and
Renewables—Smart Energy
TERIN-SEN
40127 Bologna
Italy

*Editorial Office*
MDPI
St. Alban-Anlage 66
4052 Basel, Switzerland

This is a reprint of articles from the Special Issue published online in the open access journal *Energies* (ISSN 1996-1073) (available at: https://www.mdpi.com/journal/energies/special_issues/ EERA_JPSC_2021).

For citation purposes, cite each article independently as indicated on the article page online and as indicated below:

LastName, A.A.; LastName, B.B.; LastName, C.C. Article Title. *Journal Name* **Year**, *Volume Number*, Page Range.

ISBN 978-3-0365-4913-2 (Hbk)
ISBN 978-3-0365-4914-9 (PDF)

# Contents

# About the Editor

**Paola Clerici Maestosi**

Paola Clerici Maestosi Dr., Ph.D., has been a senior scientist at ENEA Energy Technologies and Renewables Department, Smart Urban Division, since 2010. She was a practitioner from 1994 to 2000 at University of Tor Vergata, Real Estate Technical Unit, where she managed the complexity of construction and in-exercise activities at building and district levels. Due to this extensive experience and her PhD, she taught Building and Real Estate Management in Sapienza, Faculty of Architecture in Rome, from 2000 to 2007. In 2003, she supported Rome Municipality, Metropolitan Projects Department in activities related to district dimensions. From 2010 till now she worked at ENEA as a researcher in the area of Smart Cities and Positive Energy District, providing expertise in the development of smart and sustainable cities and energy communities, energy shift, energy landscape design and energy potential mapping.

For well-known attitude in managing technical complexity, multi and transdisciplinary cooperation and multi-actors involvement she is the ENEA TERIN SEN delegate in EERA Joint Program on Smart Cities (since 2010), then national delegate in Joint Programming Initiative Urban Europe since 2012 (vice-chair from 2017), national referent for EERA IWG SET Plan Action 3.2 and, last but not least, National Scientific Expert (MUR Ministry of University and Research) for DUT Driving Urban Transition Partnership (2020). In the framework of EERA Joint Program on Smart Cities she supports the creation of the EERA Joint Program on Smart Cities Special Issues Series.

EERA Special Issues Series already published:

1- 2018: European Pathways for the Smart Cities to come

2- 2019: Tools, technologies and systems integration for the Smart and Sustainable Cities to come

3- 2020: Smart Cities and Positive Energy Districts: Urban Perspectives in 2020

Actually, call for papers for 5th special issue Smart Cities and Positive Energy Districts: Urban Perspectives in 2022 has been already open and it will close mid October 2022. The EERA JPSC Special Issues Series is an EERA JPSC asset

# Preface to "Smart Cities and Positive Energy Districts: Urban Perspectives in 2021"

Cities and urban areas are the nexus for the transformations required if the EU is to achieve the targets of the European Green Deal, to fulfil commitments related to the UN's Agenda 2030 Sustainable Development Goals (SDGs), UN-Habitat's New Urban Agenda, the Urban Agenda for the EU, the Paris Agreement and provide support for the New European Bauhaus movement as well as EC Mission 100 Climate and Neutral Cities.

The urban nexus challenge is primarily characterized by the need for integrated approaches and to translate research and innovation results to actions as urban governance is still too fragmented and siloed; therefore, support for cities across levels of government is often insufficiently coordinated as public authorities often lack knowledge.

Since its creation in 2010, the EERA Joint Program on Smart Cities has accumulated important knowledge on specific topics of the program.

This includes Smart Cities and Positive Energy Districts, not only a crucial topic tackled by the EERA Joint Program on Smart Cities Workplan but also by Horizon Europe, and/or on a national level, focusing on innovative solutions based on interdisciplinary approaches, which are needed to face the highly complex structure of the transition towards sustainable urban areas.

Thanks to the most prominent national and international RD&I programs, many case studies, best practice and success cases have been completed, are ongoing, or are in their planning stage.

Accordingly, this Special Issue is promoted to support the publication of the most promising research and innovation projects established by EERA JPSC partners in recent year so as to spotlight EERA JPSC, which is not only a platform for prominent voices in the research area in Europe capable of highlighting different points of view and a variety of solutions, but also highlights selected effective interventions to support cities in their efforts to become climate neutral.

The Special Issues Series Coordinator, Paola Clerici Maestosi, conceived the Series in 2018, at a time where the need for knowledge about Smart City projects was high, thus supporting the idea to collaborate across countries thanks to the EERA JPSC network. Between 2018 and 2021, more than 70 papers from more than 10 European countries were published.

**Paola Clerici Maestosi**
*Editor*

MDPI

*Editorial*

# Smart Cities and Positive Energy Districts: Urban Perspectives in 2021

Paola Clerici Maestosi

ENEA—Italian National Agency for New Technologies, Energy and Sustainable Economic Development, Department Energy Technologies and Renewables TERIN-SEN Smart Energy Networks, 00196 Rome, Italy; paola.clerici@enea.it

**Abstract:** This Special Issue of *Energies*, "Smart Cities and Positive Energy Districts: Urban Perspectives in 2021", introduces contemporary research on Smart Cities and on Positive Energy Districts. The present Special Issue, namely the fourth Special Issue, Smart Cities and Positive Energy Districts: Urban Perspectives in 2021, has been dedicated to tools, technologies, and system integration for Smart Cities and for Positive Energy Districts. The topic highlights the variety of research within this field, including research on: tools facilitating the evaluation of Sustainable Plus Energy neighborhoods, of enabling-technologies and procedures for Positive Energy Districts, and of multicriteria assessment for the identification of Positive Energy Districts; system integration related to optimized energy and air quality management in the COVID-19 scenario; system integration upgrading existing residential areas to the status of Positive Energy Districts; and renovation models for large scale actions.

**Citation:** Clerici Maestosi, P. Smart Cities and Positive Energy Districts: Urban Perspectives in 2021. *Energies* **2022**, *15*, 2168. https://doi.org/10.3390/en15062168

Received: 6 January 2022
Accepted: 20 January 2022
Published: 16 March 2022

**Publisher's Note:** MDPI stays neutral with regard to jurisdictional claims in published maps and institutional affiliations.

## 1. Introduction

The Green Deal, Horizon Europe and the EU Urban Agenda focus on fair, green and digital transitions for urban areas, while the New European Bauhaus highlights the role of design and culture within cities; thus, the EU is shaping sustainable and livable futures as we approach a decisive moment for international efforts to tackle the climate crisis, as well as the pandemic crisis, which are the great challenges of our times.

The last few years have been characterized by a major technological boom that has seen the arrival of so-called "disruptive technologies", innovative technologies which are enabling tools for radical and positive changes to understand, plan, manage and innovate districts and urban areas. The number of countries that have pledged to reach net-zero emissions by the mid-century, or soon after, continues to grow; however, so do global greenhouse gas emissions. Thus, a total transformation of the energy system itself, and its use within urban areas, is needed.

The clean energy transition, as well as the sustainable urban area transition, must be fair, inclusive and leave nobody behind. Both transitions are for and about people. Citizens must be active participants in the entire process, making them feel part of the transition and not simply subject to it.

Technological solutions, especially digital ones, have now changed the approach to manage energy in cities. This will make it possible, in the not-too-distant future, to innovate and administer cities and services to citizens based on an understanding of their real critical issues, needs, particularities, vocations, possibilities and capabilities. The big change will not only be technological but—above all—cultural, as public administrations, citizens, economic operators, governance and business models, and market players will transform their roles within the city.

This means that, to benefit from the opportunities offered by the current technological revolution, a cultural shift must be promoted. Reaching net zero by 2050 requires further rapid deployment of available technologies, as well as a widespread use of technologies that are not on the market yet. Moreover, of particular importance is the development of national enabling conditions and conducive policy frameworks that support cities in reaching climate neutrality. Currently, urban governance is fragmented and siloed and support for cities across levels of government is often insufficiently coordinated; moreover, public authorities often lack the right skills, expertise, technical and financial resources, as well as effective intervention portfolios to support cities in their efforts to become climate neutral.

Thus, it is necessary not only to deploy available technologies or the widespread use of technologies that are not on the market yet, but also to support national and regional authorities for a transformative national change processes, by improving their multi-level governance and shaping national ecosystems for urban climate neutrality transitions.

Given this, between 2020–2027, Horizon Europe will support the clean energy transition (CET Partnership), as well as the creation of Positive Energy Districts (DUT Partnerships), which include functions such as energy efficiency (linked with Smart Cities' enabling-technologies), energy flexibility, and energy production (linked with the biggest innovation opportunity deals with advanced batteries, hydrogen electrolyzers, direct air capture and storage).

Since 2018, the EERA Joint Programme on Smart Cities has promoted and published the most promising papers on tools, technologies, and system integration for Smart Cities and Positive Energy Districts, supporting the EERA JPSC Special Issues Series with a dedicated Scientific Board and appointed Scientific Board Coordinator, Paola Clerici Maestosi.

The EERA JPSC Special Issues Series consists of:

2018—First Special Issue: European Pathways for the Smart Cities to come, https://doi.org/10.13128/Techne-2356;

2019—Second Special Issue: Tools, technologies and system integration for the Smart and Sustainable Cities to come, https://doi.org/10.5278/ijsepm.3515

2020—Third Special Issue: Smart Cities and Positive Energy Districts: Urban Perspectives in 2020, *Energies* 2021, *14*, 2351; https://doi.org/10.3390/en14092351

The present Special Issue, namely fourth Special Issue, Smart Cities and Positive Energy Districts: Urban Perspectives in 2021, has been dedicated to:

- Tools, technologies and system integration for Smart Cities;
- Tools, technologies and system integration for positive energy districts.

## 2. Published Papers Highlights

This Editorial article provides a summary of the Special Issue of *Energies*, covering the published papers [1–8] which address several of the topics mentioned in the Introduction. Table 1 identifies the most relevant topics in each published paper.

As shown in Table 1, most of the publications focus on Positive Energy Districts (6) while only two focus on Smart Cities.

The eight articles have been selected after a peer review process and we are thankful to all forty-four authors from several countries (in alphabetic order: Austria, Italy, The Netherlands, Poland, Portugal and Spain) for their contribution to the Special Issue.

In work [1] Giuseppe Anastasi et al. l revisit the current energy/environment management strategies of smart buildings and present some experimental activities carried out in the classrooms of the University of Pisa, which are used to support the proposed methodologies with an overview of the sensors used for monitoring actions, as well as the logic of the control system. Moreover, some experimental results obtained in pre-pandemic conditions, in a building of the University of Pisa, are given. The conclusion of the article drives in the direction that new approaches for designing and controlling the operation of HVAC systems are necessary and urgent to make indoor spaces safe and comfortable, without compromising energy efficiency. The installation of instruments for indoor air quality monitoring is paramount, together with the implementation of building management and control systems. Knowledge of the occupancy profile—obtained directly by means of cameras or indirectly by means of measurements of air quality parameters—represents a step forward compared to current design and operation procedures suggested by technical standards. Interaction with the occupants through active participation is a relevant element of the methodology. The argument specifically developed is that the classic topic of the optimization of a Building Management System (BMS) is becoming even more relevant in a COVID-19 context.

**Table 1.** Topics covered in each publication.

| NO | Title | Tools, Technologies and System Integration for Smart Cities | Tools, Technologies and System for Positive Energy Districts |
|---|---|---|---|
| 1 | Optimized Energy and Air Quality Management of Shared Smart Buildings in the COVID-19 Scenario | ■ | |
| 2 | PEDRERA. Positive Energy District Renovation Model for Large Scale Actions | | ■ |
| 3 | Combining Sufficiency, Efficiency and Flexibility to Achieve Positive Energy Districts Targets | | ■ |
| 4 | Analysis and Evaluation of the Feasibility of Positive Energy Districts in Selected Urban Typologies in Vienna Using a Bottom-Up District Energy Modelling Approach | | ■ |
| 5 | An Evaluation Framework for Sustainable Plus Energy Neighbourhoods: Moving Beyond the Traditional Building Energy Assessment | ■ | |
| 6 | Possibilities of Upgrading Warsaw Existing Residential Area to Status of Positive Energy Districts | | ■ |
| 7 | A GIS-Based Multicriteria Assessment for Identification of Positive Energy Districts Boundary in Cities | | ■ |
| 8 | Positive Energy Districts and Energy Efficiency in Buildings: An Innovative Technical Communication Sheet to Facilitate Policy Officers' Understanding to Enable Technologies and Procedure | | ■ |

Article [2] by P. Civiero, J. Pascual, J. Arcas Abella, A. Bilbao Figuero and J. Salom provides a view of the ongoing PEDRERA project, whose main scope is to design a district simulation model able to set and analyze a reliable prediction of potential business scenarios on large scale retrofitting actions, and to evaluate the overall co-benefits resulting from the renovation process of a cluster of buildings. According to this purpose, and to a Positive Energy Districts (PEDs) approach, the model combines systemized data—at both building and district scale—from multiple sources and domains. Once implemented, the PEDRERA tool will facilitate the engagement of multiple stakeholders involved in the building renovation programs to make effective and well-informed decisions from a cluster of georeferenced buildings. Preliminary results obtained by the ongoing PEDRERA project refer to the definition of the conceptual framework of the model, and in regards to: (a) data source aggregation according to the four domains described above; (b) input required for the KPI calculations which can be assumed to assess different "scopes". Aligned with this vision, the model is powered by the integration of the processed input for the calculation of the most relevant KPI (outputs) algorithms, according to each process phase and stakeholder's profile.

Contributions [3] proposed by Silvia Erba and Lorenzo Pagliano provide a matrix of interactions between building and district design for use by building designers and city planners; they compare possible scenarios implementing different strategies at the building and urban levels in a case study, in order to evaluate the effect of the proposed integrated approach on the energy balance at yearly and seasonal time scales, and on land take. The main findings rely on the energy sufficiency enablers.

Analysis and Evaluation of the feasibility of Positive Energy Districts in selected urban typologies in Vienna is presented in article [4] by Hans-Martin Neumann et al., which investigates the potential of selected urban typologies in Vienna to reach the state of Positive Energy Districts (PED) by achieving a positive annual energy balance. Considering relevant urban typologies in different construction periods, the analysis focused on converting the allocated building stocks into PED by employing comprehensive thermal refurbishment and energy efficiency measures, electrification of end-uses and fuel switching, exploitation of local renewable energy potential, and flexible interaction with the regional energy system. The results reveal that a detached housing district can achieve a positive annual energy balance (for heat and power) of 110%, due to the fact that there are sufficient surfaces (roofs, facades, open land) available for the production of local renewable energy; conversely, the remaining typologies fail to achieve the criteria, with an annual balance ranking of between 61% and 97%, showing additional margins for improvement to meet the PED conditions.

Jaume Salom et al. [5] contribute to the on-going debate about the definition of the notion of a Sustainable Plus Energy Neighborhood (SPEN), which highlights the need to consider mutual interaction between the built environment, the inhabitants and nature. Through a multidimensional analysis to address complexity in neighborhoods, this paper outlines an assessment framework for the performance evaluation of SPEN. The selection of the main assessed categories and Key Performance Indicators (KPIs) have been based on a holistic and comprehensive methodology which highlights the multiple dimensions of sustainability in the built environment. The contents of the paper are based on the work developed in the syn.ikia project with extended details on the methodology applied, revised definitions and concise and synthetic presentation of the metrics. As result of the application of the methodology described, five KPI categories were identified, and are defined as: Energy and Environmental, which addresses overall energy and environmental performance, matching factors between load and on-site renewable generation and grid interaction; Economic, addressing capital costs and operational costs; Indoor Environmental Quality (IEQ), addressing thermal and visual comfort, as well as indoor air quality; Social, which addresses the aspects of equity, community and human outcomes; and Smartness and Flexibility, addressing the ability to be smartly managed. This paper presents two main contributions to the existing literature. The first is to put forth a consistent definition of a Sustainable Plus Energy Neighborhood (SPEN), while the second major contribution is to present an evaluation framework for the assessment of SPENs.

In [6], Hanna Jędrzejuk and Dorota Chwieduk analyze the possibilities for the refurbishment of Warsaw's residential buildings towards the standards of a Positive Energy District. The paper refers to the theoretical potential for the use of renewable energies, which appears to be a rational solution only if there is a reduction in energy demand for traditional methods which, in this case, means reducing the energy demand of existing buildings or erecting new ones in accordance with the latest energy and environmental standards. Potential barriers to the implementation of renewable energy technologies and achieving the status of a Smart City with some positive energy districts should be identified and mentioned.

Beril Alpagut et al., paper [7], focuses on a flexible GIS-based Multicriteria assessment method that identifies the most suitable areas to reach an annual positive non-renewable energy balance. For that purpose, a GIS-based tool is developed to indicate the boundary from an energy perspective harmonized with urban design and land-use planning. The method emphasizes evaluation through economic, social, political, legal, environmental, and technical criteria, and the results present the suitability of areas at macro and micro scales. The current study outlines macro-scale analyses in six European cities that represent Follower Cities under the Making-City H2020 project.

Tiziana Ferrante and Teresa Villani [8] refer to an innovative technical communication sheet to facilitate policy officers' understanding about enabling-technologies and procedure to support the transition to a Positive Energy District. Based on the results of ENEA national research [Sustainable Urban Transition; pp. 240–241; https://doi.org/10.30448/UNI.916.50733], which individuated more than 100 key indicators for six areas of investigations to describe PED experiences, T. Ferrante and T. Villani elaborated an innovative technical communication sheet which describes—in a visual, effective and easy way—energy efficiency key indicators and the related implementation process. As key indicators refer both to technological solutions and the execution process, two types of technical communication sheets have been created; the first refers to technological solutions and highlights building system components related to energy efficiency, while the second is about the implementation phase. The article presents two case studies (Milan, Nido in Feltrinelli; Florence, Scuola Materna Capuana) where technical communication sheets have been applied to describe the rehabilitation project supporting the comparison between different technological solutions and implementation procedures. The results presented in this paper seem interesting and in line with ongoing European debate, as there is a need for effective communication on PED case studies highlighting information related to the strategies and solutions implemented by the municipalities. Authors affirm that effective communication on findings related to PED will encourage and facilitate synergies among urban ecosystem stakeholders, by activating a virtuous communication processes and improving the understanding of actions to support the PED transition.

**Conflicts of Interest:** The author declares no conflict of interest.

# References


1. Anastasi, G.; Bartoli, C.; Conti, P.; Crisostomi, E.; Franco, A.; Saponara, S.; Testi, D.; Thomopulos, D.; Vallati, C. Optimized Energy and Air Quality Management of Shared Smart Buildings in the COVID-19 Scenario. *Energies* **2021**, *14*, 2124. [CrossRef]
2. Civiero, P.; Pascual, J.; Arcas Abella, J.; Bilbao Figuero, A.; Salom, J. PEDRERA. Positive Energy District Renovation Model for Large Scale Actions. *Energies* **2021**, *14*, 2833. [CrossRef]
3. Erba, S.; Pagliano, L. Combining Sufficiency, Efficiency and Flexibility to Achieve Positive Energy Districts Targets. *Energies* **2021**, *14*, 4697. [CrossRef]
4. Neumann, H.-M.; Hainoun, A.; Stollnberger, R.; Etminan, G.; Schaffler, V. Analysis and Evaluation of the Feasibility of Positive Energy Districts in Selected Urban Typologies in Vienna Using a Bottom-Up District Energy Modelling Approach. *Energies* **2021**, *14*, 4449. [CrossRef]
5. Salom, J.; Tamm, M.; Andresen, I.; Cali, D.; Magyari, Á.; Bukovszki, V.; Balázs, R.; Dorizas, P.V.; Toth, Z.; Mafé, C.; et al. An Evaluation Framework for Sustainable Plus Energy Neighbourhoods: Moving Beyond the Traditional Building Energy Assessment. *Energies* **2021**, *14*, 4314. [CrossRef]
6. Jędrzejuk, H.; Chwieduk, D. Possibilities of Upgrading Warsaw Existing Residential Area to Status of Positive Energy Districts. *Energies* **2021**, *14*, 5984. [CrossRef]
7. Alpagut, B.; Lopez Romo, A.; Hernández, P.; Tabanoğlu, O.; Hermoso Martinez, N. A GIS-Based Multicriteria Assessment for Identification of Positive Energy Districts Boundary in Cities. *Energies* **2021**, *14*, 7517. [CrossRef]
8. Ferrante, T.; Villani, T. Positive Energy Districts and Energy Efficiency in Buildings: An Innovative Technical Communication Sheet to Facilitate Policy Officers' Understanding to Enable Technologies and Procedure. *Energies* **2021**, *14*, 8551. [CrossRef]

*Article*

# Analysis and Evaluation of the Feasibility of Positive Energy Districts in Selected Urban Typologies in Vienna Using a Bottom-Up District Energy Modelling Approach

**Hans-Martin Neumann [1],*, Ali Hainoun [1], Romana Stollnberger [1], Ghazal Etminan [1] and Volker Schaffler [2]**

[1] AIT Austrian Institute of Technology GmbH, 1210 Vienna, Austria; ali.hainoun@ait.ac.at (A.H.); romana.stollnberger@ait.ac.at (R.S.); ghazal.etminan@ait.ac.at (G.E.)

[2] Federal Ministry for Climate Action, Environment, Energy, Mobility, Innovation and Technology (BMK) of Austria, 1030 Vienna, Austria; volker.schaffler@bmk.gv.at

* Correspondence: hans-martin.neumann@ait.ac.at; Tel.: +43-50550-6071

**Abstract:** This article investigates the potential of selected urban typologies in Vienna to reach the state of Positive Energy Districts (PED) by achieving a positive annual energy balance. It follows the EU initiative for implementing at least 100 PED in Europe by 2025. Four urban typologies have been assessed using the bottom-up energy modelling tool MAPED that enables a simplified energy demand-supply analysis at the district scale. Considering relevant urban typologies in different construction periods, the analysis focused on converting the allocated building stocks into PED by employing comprehensive thermal refurbishment and energy efficiency measures, electrification of end-uses and fuel switching, exploitation of local renewable energy potential, and flexible interaction with the regional energy system. The results reveal that a detached housing district can achieve a positive annual energy balance (for heat and power) of 110% due to the fact that there are sufficient surfaces (roofs, facades, open land) available for the production of local renewable energy, whereas the remaining typologies fail to achieve the criteria with an annual balance ranking between 61% and 97%, showing additional margins for improvement to meet the PED conditions. The presented concept offers a practical approach to investigate the PED suitability of urban typologies. It will help the Austrian Ministry for Climate Action and Environment to identify appropriate strategies for the refurbishment of existing urban areas towards the PED standard.

**Keywords:** Positive Energy Districts; urban typology; energy modelling; energy and climate goals; energy flexibility; sustainable urban development

**Citation:** Neumann, H.-M.; Hainoun, A.; Stollnberger, R.; Etminan, G.; Schaffler, V. Analysis and Evaluation of the Feasibility of Positive Energy Districts in Selected Urban Typologies in Vienna Using a Bottom-Up District Energy Modelling Approach. *Energies* **2021**, *14*, 4449. https://doi.org/10.3390/en14154449

Academic Editors: Paola Clerici Maestosi and Luisa F. Cabeza

Received: 1 May 2021
Accepted: 15 July 2021
Published: 23 July 2021

**Publisher's Note:** MDPI stays neutral with regard to jurisdictional claims in published maps and institutional affiliations.

## 1. Introduction

It has become apparent that sustainable urban development can only be achieved through a significant change in the way we build and manage our urban spaces. Transforming urban energy system is a key driver of the aspired development to make cities and human settlements inclusive, sustainable, and resilient, as elaborated in Goal 11 of the UN Sustainable Development Goals (SDGs) [1]. Within this effort, Positive Energy Districts (PEDs) represents an innovative concept for the development of urban districts and neighbourhoods.

PEDs refer to urban neighbourhoods with the ability to achieve a positive energy balance on an annual base within its given boundary. This means that the cumulative annual energy provided within the district boundary must exceed its annual own demand and compensate for any external energy supply. Hereafter, defining the system boundary of a PED is crucial for achieving an annual positive energy balance because of internal energy consumption and local renewable energy production. Such boundaries might refer to geographical, functional, or/and virtual domains [2].

The Joint Programming Initiative Urban Europe (JPI-UE) proposes the following definition in its publication, "Framework Definition for Positive Energy Districts and Neighbourhoods" [3]:

"Positive Energy Districts are energy-efficient and energy-flexible urban areas which produce net zero greenhouse gas emissions and actively manage an annual local or regional surplus production of renewable energy. They require integration of different systems and infrastructures and interaction between buildings, the users and the regional energy, mobility and ICT systems, while optimising the liveability of the urban environment in line with social, economic and environmental sustainability".

This definition builds also on previous contributions and ongoing discussions around the realisation and deployment of PEDs as originally highlighted by Temporary Working Group (TWG) 3.2 of SET-Plan Action 3.2 on implementation plan of PEDs [2].

The resulting definition highlights three pillars for realising PEDs from the viewpoint of sustainable energy system (Figure 1):

1. High level of energy efficiency: to keep district annual energy consumption as low as achievable.
2. Local/regional renewable energy supply:
3. Optimised and flexible energy system: to optimised interaction with the neighbourhood energy system and manage consumption and storage capacities on demand.

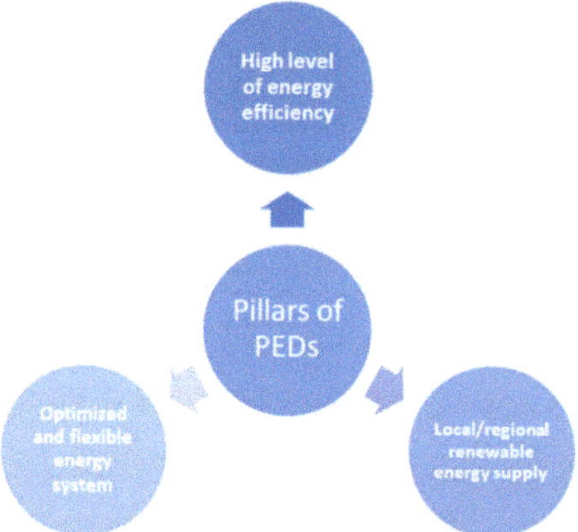


**Figure 1.** Functions and pillars of Positive Energy Districts (PEDs) within the urban energy system (Source: AIT).

Following the above definition PEDs seeks to optimise the three functions towards climate neutrality and energy surplus considering the guiding principles of quality of life, inclusiveness and sustainability. It has been recognised that the contribution of various key stakeholders and enablers is essential to enable the realisation and deployment of PEDs covering:

urban governance and regulatory framework,
Engagement of citizen and need-owners,
Integration of urban and energy planning for a sustainable and resilient PEDs
Employment
ICT and integrated energy solutions, e.g., sector coupling and EV.

## 2. The Positive Energy Districts (PEDs) Policy Context

The main driver for PEDs is currently the climate and energy policy of the European Union (EU) and its member states. The EU is speeding up innovation in clean energy and calls for decarbonization of the EU building stock by 2050 [4]. Such a transformation requires innovative technological solutions (with a focus on the integration of energy systems), in addition to regulation, financing, governance, new business models, and other associated socio-economic issues. Several European initiatives have been taken in this direction, aimed at achieving the EU long-term energy and climate goals; among them are the SET-Plan Action 3.2 "Smart Cities and Communities" [5] and the recently adopted European Green Deal to reach climate-neutrality by 2050 [6].

The prominent EU initiative under SET-Plan Action 3.2 SCC considers PEDs as a driver of sustainable urbanisation. In its declaration of intent published in 2016, it seeks to "make Europe a global role model in integrated, innovative solutions for the planning, deployment and replication of Positive Energy Districts with the aim by 2025 to have at least 100 Positive Energy Districts synergistically connected to the energy system in Europe and a strong export of related technologies" [5]. This initiative is the result of extensive consultations with several stakeholders, including European Innovation Partnership on Smart Cities and Communities, Covenant of Mayors, EERA Joint Programme on Smart Cities, Joint Program Initiative Urban Europe, EU SCIS, ERA-NET on Smart cities and Communities, beside public consultation. The initiative stresses that PEDs raise the quality of life in European cities, contribute to reach the COP21 targets, and enhance European capacities and knowledge to become a global role model. Following this initiative, the temporary technical working group 3.2 (TWG 3.2) was established. Chaired by national representatives from Austria, TWG 3.2 developed in a joint effort a pathway towards PED in Europe, including a technology roadmap. In addition, it specified commitment for planning and implementation actions [2]. The outcomes of this effort resulted in proposing an integrated approach to tackle the interdisciplinary challenges of PEDs covering technological, economic, financial, legal, and regulatory aspects within an urban perspective. Moreover, it recognised the crucial role of cities on the way to realise PEDs together with the vital contribution of key stakeholders from research, industry, real state, and funding and financing, beside other fields.

Based on this initiative, the transnational Joint Programming Initiative Urban Europe (JPI-UE) has been working to provide a programme management structure for PEDs research activities. It aligns research efforts with cities' needs and their apprehended future sustainable development goals, including the deployment of PEDs. The JPI programme seeks, for its implementation, the contribution of stakeholders of city authorities, research organisation, industry, energy suppliers, and citizens' organisations. [3].

Considering the attractiveness of the PEDs concept for sustainable urban development, numerous initiatives have been taken to develop solutions, roadmaps, and business models for planning and implementing PEDs [7,8]. Such initiatives benefit from the wealth of experiences gained in the construction of positive energy buildings (PEBs) that form the building blocks of future PEDs, leveraging innovative technologies for building, integrating, and managing buildings within an integrated neighbourhood energy system.

The IEA Annex 83 "Positive Energy Districts" [9], a research network under the auspices of the International Energy Agency (IEA), is currently documenting the international state of the art. Besides several European Universities and research organizations, it involves participants from Canada, Australia, Japan, and China (Hong Kong). In the United States of America, for example, researchers from the National Renewable Energy Laboratory (ENREL) have developed a definition for a "Zero Net Energy Community" [10]. NREL is also involved in the Smart City project and Zero Energy District "Peña Station" in Denver [11]. These initiatives show that the interest in PED is growing rapidly, not only in Europe, but also globally.

However, PEDs are still in the early stage of their introduction with a significant need to overcome a multitude of challenges spanning across technological, financial, environmental, societal, and regulatory domains. This stems from the fact that a PED is not just an energy standard, but rather an innovative concept to promote the sustainable development of urban energy systems on a district scale with significant impact on the development of our future cities, which are committed to a sustainable and low-carbon pathway to ensure high viability and affordability of urban services for all residents.

### 3. From Zero Energy Buildings to Positive Energy Districts

The concept of PEDs is related to the concept of Near Zero Energy Buildings (NZEB) and Zero Energy Buildings, for which several concepts have been developed and demonstrated worldwide. Not only do Net Zero Energy Buildings (NZEBs) and Zero Energy Buildings (ZEB) not consume energy, but they also generate renewable energy onsite. This allows for a high share of self-consumption and thus a reduced carbon footprint [12].

Many studies have analysed how buildings can be designed, built, maintained, and refurbished to become NZEBs or even ZEBs [13]. The findings of these studies provide valuable insights into the development of PEDs. However, there are significant differences in the challenges that need to be taken when an NZEB is planned and a PED is developed. However, framework conditions for the planning and implementation of a Zero Energy Building and Positive Energy Districts differ significantly:

Zero Energy Buildings are usually new buildings, planned and built by one developer. The architects, engineers, and other technical experts are involved in the project work on behalf of this developer. This means that plans and other data are exchanged freely within the planning team. After the realization, the building is usually operated and maintained by a single building operator on behalf of the owner. Positive Energy Districts, however, usually consist of already existing buildings, with new buildings as infills. Typically, the buildings have several owners and operators. Information on energy consumption and building technology tends to be incomplete. The transformation of a neighbourhood into PED takes longer than the planning and construction of a ZEB, as not all necessary measures can be implemented at the same time. During the planning and implementation of PEDs, a multitude of actors need to be involved, including not only the building owners, but also tenants, energy utilities, and several branches of the city administration. For the operation phase, a multi-party energy management system and contractual arrangements for the exchange of energy (e.g., in the form of a Renewable Energy Community) must be set up. Due to these structural differences, PEDs require different planning approaches and tools than ZEBs. This applies to the pre-assessment, planning, and monitoring and evaluation phases. In this article, we will present a planning method that allows to pre-assess which urban neighbourhoods have the potential to become a PED, based on urban space types.

Urban space types have been used in several studies in urban energy planning. Everding and Kloos [14] developed prototypes for solar urban neighbourhoods, relating to 14 urban spaces found in many cities. Genske, Jödecke, and Ruff [15] developed a tool to identify the potential of renewable energy supply within different types of typologies of urban neighbourhoods and open spaces. The tool assesses the potential of not only roofs in generating renewable energy, but also façades as well as the immediate surroundings of buildings, urban open spaces, and the urban subsoil. An updated version of this tool is described in Everding, Genske, and Ruff [16], where their energy model is discussed in detail. Another example of the use of space types to model urban energy demand and potential for renewable energy was developed by Hegger and Dettmar [17]. Their typology of urban spaces provides information on the energy and structural characteristics of typical settlement forms. In addition, they characterize green open spaces, water areas, and street spaces by energy requirements and potential. Although several typologies and related urban energy models have been developed in recent years, they have so far not been tailored for the pre-feasibility assessment of PEDs.

## 4. Approach and Methodology for Evaluating Positive Energy Districts (PEDs)

Bringing PED concepts into implementation requires conducting pre-feasibility studies that rely on mapping promising urban typologies and examine their conversion potential towards PEDs. This article presents a new approach to assess the potential of different urban typologies to reach a PED standard, based on typology, for urban neighbourhoods. The aim is to provide an easy-to-use and applicable approach to test and inspect the potential for implementing PEDs in cities and municipalities that aim towards a carbon-neutral future.

Our work is embedded within this realm and provides its scientific contribution in two-folds. First, offering a systematic approach for mapping urban typology and examining their conversion potential towards PEDs, considering the social, technological, climate and urban planning criteria of the nominated sites. Second, conducting a simplified quantitative energy assessment of the selected site to define and specify the needed measures to attain a PED with annual positive energy balance.

Building on the elaborated approach of defining and specifying PEDs, a simplified concept was developed to evaluate the suitability of different urban typologies to generate more energy than it consumes and reach the status of one of the defined PED types. For this purpose, the bottom-up modelling tool MAPED (Model for Energy Analysis of Positive Energy District) was developed. MAPED enables the user to analyse and evaluate the energy demand-supply of urban districts and additionally offers the possibility to test different scenarios and implementation measures to explore the transformation pathways towards a PED.

To retrieve a typical district's potential to become a PED, the following steps need to be followed:

- Identification of relevant urban typologies (detached housing area/single-family homes (SFH), terraced housing (TH) area, multi-family housing (MFH) area, apartment blocks (AB)) for different construction classes,
- Data collection as input for the MAPED tool using the GIS-based approach to extract area boundaries, population, built-up area, building footprints, etc.,
- Mapping and modelling the district energy system within MAPED,
- Analyses of different supply options and related conditions for implementation.

### 4.1. Selection of Different Urban Typologies

To ensure applicability in European cities, it was important to select different urban typologies that can be refurbished and built elsewhere as well. Based on Vienna´s urban neighbourhood typology that was developed by the Municipal Department for Urban Development and Planning [18], four typologies were selected (Figure 2):

1. Detached housing built 1961–1980: 91% single-family homes (SFH), 9% terraced houses (TH)/multi-family homes (MFH),
2. Dense inner-city area (Gründerzeit), built before 1919: ~100% MFH,
3. Medium dense area built between 1961 and 1980 (economic boom): 63% SFH, 5% TH, 32% MFH, non-residential use with about 7% of the gross floor area,
4. Detached housing constructed from 2006: 100% SFH.

**Figure 2.** Selected urban typologies (according to [19]).

## 4.2. Mapping of District Data

Based on the typologies, the selected areas were assigned data in a further step. The data was derived from the official buildings and dwellings register [20], mapped to statistical grid cells of 250 m× 250 m from the statistical office (Table 1). The dataset contains important data for the analyses with MAPED, such as population, buildings by type of use (e.g., residential building with number of dwellings, hotel, office building, retail, agricultural use, etc), and buildings by construction period. By means of geographic information systems (GIS) and various geodata layers, relevant parameters were extracted and calculated within the 250 m grid typology, such as the districts´ total, gross, and built areas; type, size, and height of residential and non-residential buildings (dwelling and service), and population and number of dwellings. Based on the selected typologies and the related official buildings and dwellings register, the key data needed to conduct the simplified energy demand-supply analysis for the considered districts were prepared (Table 2).

## 4.3. Short Description of the MAPED (Model for Energy Analysis of Positive Energy District) Model

MAPED is a bottom-up rapid energy assessment tool for analysing the energy demand and supply of urban districts and assessing their qualification to reach an annual positive energy balance by exploiting local RES to cover a district's electricity and heat demand. MAPED was developed by AIT, based on the proven end-use concept of the IAEA model MAED [21]. MAPED focuses on the evaluation of useful and final energy demands at the district scale, covering energy demands for residential and non-residential building, urban farming, industry, and mobility. Moreover, it offers a simplified approach to evaluate and estimate local renewable energy production to cover heat and electricity demands using photovoltaic, solar thermal energy, and heat pumps. Other local supply options like biomass, waste heat, and micro wind can be also considered, given the prevailing boundaries, topology, social acceptance, and the applied regulations (Figure 3). The MAPED approach evaluates final and useful energy demands based on the demographic, social, and technological data of the considered district and services to the social, economic, and technological factors that affect the demand for a particular fuel (this could also include urban farming and local industrial activities in case of their existence). This implies population number and growth, number of inhabitants per dwelling, number of electrical appliances used in households and services, peoples' mobility and preferences for transport modes, evolution of the efficiency of certain types of equipment, and market penetration of new technologies or energy forms. The expected future trends for these determining factors, which constitute 'scenarios', are exogenously introduced. This enables evaluation of the needed measures to convert the considered district to a PED within the given demographic, social, technical, and building types' specifications.

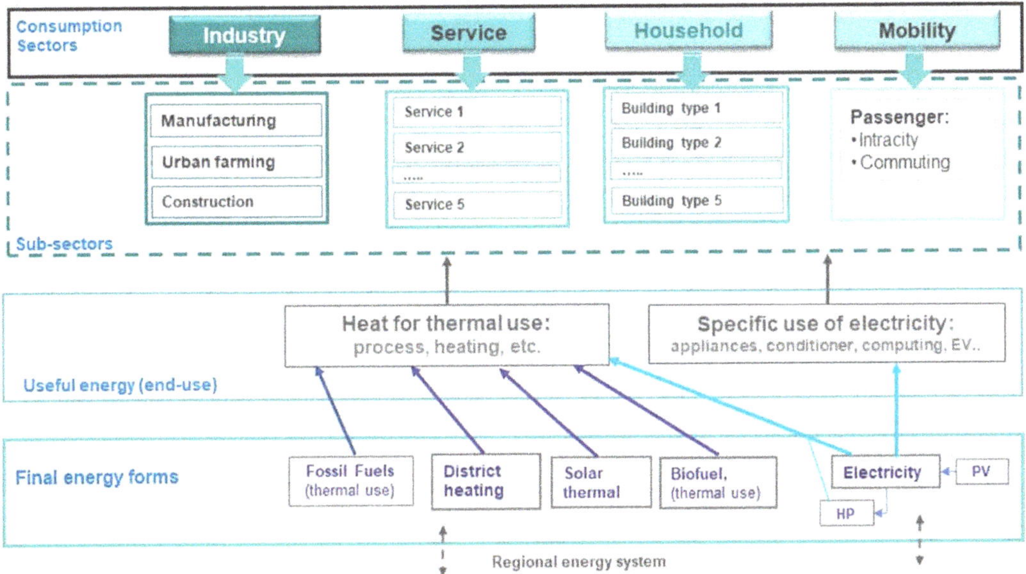


**Figure 3.** MAPED (Model for Energy Analysis of Positive Energy District) concept for the disaggregation of district energy demand, by fuel and consumption sectors (source: AIT); HP—heat pump, PV—photovoltaics.

For the analysis of energy usage in households, five types of dwellings are included, each type described by the size of the apartment/house, number of people living in each type, and energy efficiency of each type for heating and cooling. The service sector (non-residential building) is modelled based on the type of economic activities that affect the type of buildings, and their energy uses. Thus, five groups of service activities are considered, covering offices, educational institutions, shopping and commercial activities, hospitals, hotel, and restaurants. This feature allows for a realistic analysis of energy demands and enables to explore the possibility of a "Positive Energy District", given the type of building operational energy and the potential of local renewables and their technical exploitation. Based on the conducted district energy demand-supply analysis, key indicators are generated to evaluate the district self-sufficiency to cover its energy demand and achieve a positive annual energy balance.

### 4.4. District Input Data for MAPED (Model for Energy Analysis of Positive Energy District) Analysis

Following the above-described end-use approach of MAPED, several input parameters are needed to conduct a final energy demand analysis describing the current state and the conceived future development towards the set target of a PED, assuming to be reached by 2040. Beside the specified urban typology (Table 1), the input data cover the following categories:

- Demographic, social data: total population, dwelling size, person per dwelling, share of each dwelling type,
- Climate data: heat degree days (HDD) and cooling degree days (CDD),
- Technological data: average heat loss coefficient of building envelope, efficiency, and penetration rates of final fuel consumption by end-use category (e.g., share and efficiency of gas and biomass for covering space heating), share of dwelling area cooled,
- Specific energy consumption data: current final energy consumption by fuel type and end-use form for space heating and cooling, water heating and cooking, lighting, appliances,
- Potential of local renewables: estimated potential for PV on rooftop area, facades, open land area, and transport infrastructure area in the district.

**Table 1.** Base data of the selected urban typologies.

| Data Description | Detached Housing 1961–1980 | Dense Inner City before 1919 | Medium Dense 1961–1980 | Detached Housing 2006 |
|---|---|---|---|---|
| Total area [m$^2$] | 62,500 m$^2$ | 62,500 m$^2$ | 125,000 m$^2$ | 62,500 m$^2$ |
| Gross floor area [m$^2$] | 23,760 m$^2$ | 100,250 m$^2$ | 66,510 m$^2$ | 19,400 m$^2$ |
| Share of residential area | 100% | 100% | 93% | 100% |
| Share of service area | - | - | 7% | - |
| Service sector floor area | - | - | 1970 (22% nonresidential: 11% commercial and 11% hotels/restaurants) | - |
| Built-up area [%/m$^2$] | 27%/17,170 m$^2$ | 50%/31,520 m$^2$ | 19%/23,200 m$^2$ | 17%/10,420 m$^2$ |
| Traffic area [%/m$^2$] | 17%/10,330 m$^2$ | 30%/18,760 m$^2$ | 20%/24,440 m$^2$ | 16%/10,210 m$^2$ |
| Green area [%/m$^2$] | 56%/35,000 m$^2$ (private green) | - | 2%/2000 m$^2$ (public green) | 9%/5450 m$^2$ (public green) |
| Population | 312 | 2512 | 1071 | 376 |
| Number of dwellings | 154 | 1429 | 521 | 143 |
| Single-family homes [share/area in m$^2$] | 91%/150 m$^2$ | - | 63%/150 m$^2$ | 99%/150 m$^2$ |
| Building with 2 dwellings [share/area in m$^2$] | 2%/75 m$^2$ | 1%/75 m$^2$ | 5%/75 m$^2$ | 1%/75 m$^2$ |
| Building with 3 or more dwellings [share/area in m$^2$] | 7%/65 m$^2$ | 99%/65 m$^2$ | 32%/65 m$^2$ | - |

The specific energy consumption data per dwelling (and dwelling size) and by end-use activities of water heating, cooking electricity consumption of appliances were collected from the energy survey on household energy consumption of statistics in Austria [22]. The current technical state of building insulation for the considered typologies and building types refers to the standard energy performance certificate of the defined "generic building types" for Austria provided by the TABULA Typology structure of the EPISCOPE project [23,24]. To reach the status of a PED, measures specifically targeting the framework conditions and challenges for each typology must be applied. While it is difficult for dense inner areas to harvest locally available energy sources due to the limited availability of space for renewable infrastructure plants like large-scale PV on open land, it is easier for detached housing districts due to more land availability. Another limiting factor in multi-family-houses is the dependency on many tenants if some plants are to be refurbished or newly built. Table 2 presents exemplarily the data used for the first type, "Detached housing".

For future developments towards the PED 2040 target, the following assumptions have been applied:

- Adopting effective building refurbishments to reach an advanced level of space heat energy performance, according to the Austrian building standard OIB RL 6 for low energy buildings [25],
- Significant efficiency improvements in space and water heating, lighting, and appliances over the period of 2020–2040,
- Fuel switch from fossil to renewable supply with focus on electrification of end-use activities of cooking, space heating, and water heating via a heat pump (HP) beside solar thermal energy (ST),
- No biomass (BM) is considered,
- Increasing the penetration rate of HP and ST to fully cover the heat demands for space and water heating by 2040, as follows:
    - Space heating: 75% HP, 15% ST, and 10% direct electricity as a backup system for ST
    - Water heating: 63% HP, 30% ST, and 7% direct electricity as a backup system for ST
- Improvement in the Coefficient of Performance (COP) of HP by around 37% to reach 3.8 by 2040,
- Utilisation of the top local renewable energy potential with focus on PV and ST, aiming at meeting the annual electricity and heat demands from local renewables,
- Interaction with the electricity grid of the city, beside local power storage, has been assumed but not explicitly modelled,
- Increasing the share of dwellings requiring cooling from 5% to 20% over the period of 2020–2040
- No change in population number, person per dwelling, Heating Degree Days (HDD), and Cooling Degree Days (CDD).

**Table 2.** Annual key parameter for modelling the energy demand of the detached housing district.

| Data Description | Current State 2020 | PED-Target 2040 |
| --- | --- | --- |
| Population | 312 | 312 |
| HDD/CDD | 2919/857 | 2919/857 |
| Heat loss coefficient $(W/m^2K)$/ EPC [1] $(kWh/m^2a)$ | 2.06/144.3 | 1.0/70 |
| Water heating (kWh/cap) | 1054 | 843 |
| Cooking (kWh/dw) | 500 | 500 |
| Lighting (kWh/dw) | 365 | 292 |
| Appliances, non-shiftable (kWh/dw) | 838 | 670 |
| Appliances, shiftable (kWh/dw) | 1146 | 917 |
| Penetration of energy forms into SH: elec./district heating/fossil/BM/ST | 12.8%/32%/50.2%/2.5%/2.5% | 85%/-/-/-/15% |
| Penetration of energy forms into WH: elec./district heating/fossil/BM/ST | 18.9%/35.6%/36.8%/0.4%/8.3% | 35%/-/-/65% |
| COP of HP | 2.5 | 3.8 |

[1] Energy Performance Certificate; HDD—Heating Degree Days, CDD—Cooling Degree Days; SH—space heating; BM—biomass; ST—solar thermal; WH—Water heating; COP—Coefficient of Performance, HP—heat pump.

## 5. Results and Discussion

Following the above-presented concept and the compiled input data, a detailed energy demand-supply analysis has been conducted for each of the four considered districts. The following sections demonstrate the results for the typology of the first district type (Detached Housing District). The remaining three typologies are addressed in a similar way.

### 5.1. Energy Demand

The adopted transformation measures on the demand side in terms of efficiency improvement and electrification of end-uses will boost the overall district energy efficiency by 66.2%, resulting in annual district final energy demand reduction from 3.94 GWh to

1.33 GWh. The applied fuel switching and electrification for end-use will lift the share of electricity from the current 20% to 72% of the total final demand of the projected PED. The resulting average annual final energy demand per dwelling will decrease from the current 25,552 kWh/Dw (dwelling) to 8640 kWh/Dw of the targeted PED, corresponding to 179 kWh/m$^2$ and 61 kWh/m$^2$, respectively (Figure 4). The decrease of space heating from the current 80% to 50% for PED state is remarkable, beside the increased share of appliances from 7% to 18%. The achieved reduction in space heat final energy demand, which dominates and drives the desired transition to PED, is the direct result of two combined measures, namely effective building refurbishments and the shift to highly efficient HP—its share is assumed to rise from the current 3.4% to 70%.

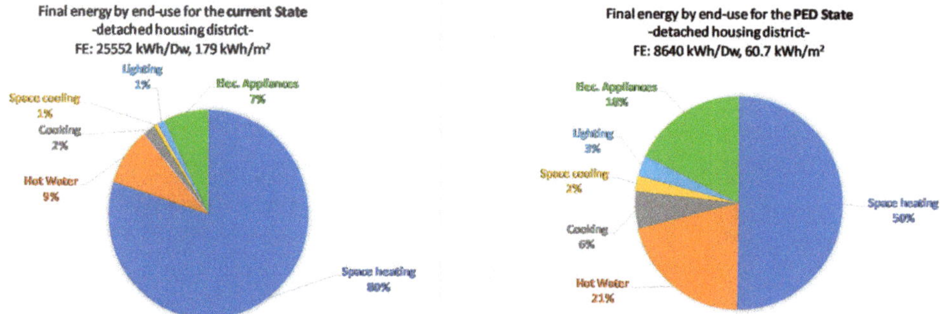


**Figure 4.** Distribution of final energy demand by end-use activities for the expected PED, compared to the current state for the considered detached housing district (source: AIT).

The specific final energy for space heat demand is estimated to drop from 144 kWh/m$^2$ to 30.6 kWh/m$^2$ and the useful space heat demand (corresponding to the EPC) from 128.8 to 70 kWh/m$^2$, following the adopted effective building refurbishment. The results indicate a transition from externally provided supply in terms of fossil fuel and district heating to local supply of HP and ST.

*5.2. Energy Supply*

The observed transformation will be enabled mainly through the electrification of final energy, enabled by the local renewables supply of PV, besides the contribution of ST to cover part of the hot water and space heating. On an annual basis, around 72% of the district final energy demand will be provided by PV and 28% by ST (Figure 5). Around 76% of ST will be devoted to HW and the remaining 24% to SH. Local PV-generation covers the remaining HW demand, main part of SH (76% via HP), and all electricity demands for appliances, lighting, space cooling (SC), and cooking. Moreover, it is assumed that the district will interact with regional electric and heat networks in the neighbourhood to account for the needed flexibility to compensate for energy deficit and surplus over various periods of the year. The technical potential of PV depends on available areas of rooftops, south facades, open land, and transport infrastructure. For a realistic harnessing of the local renewable energy of the detached housing district, the following combination was adopted after intensive consultation with building developers:

- 40% of the roof top area is used for PV panels: intense consultations with real estate developers reveals that no more than 60% of the roof area can be utilised for the installation of PV or solar thermal. The adopted figures in this analysis is based on experts' recommendations.
- 5% of the roof top area is used for ST collectors,
- 10% of the south façade is used for PV panels.

The resulting electricity generation density (yield per square meter) for PV amounts to 144.7 kWh/m$^2$, comparable to the documented average value of 153.8 kWh/m$^2$ for Vienna (MA20, 2018). For solar thermal, the resulting thermal generation density is around 447.5 kWh/m$^2$, which is close to the highest confidence value specified for Vienna in the range of 450–600 kWh/m$^2$ for the low temperature heat [26]. The resulting annual output is 1.054 GWh for PV and 0.384 GWh for ST.

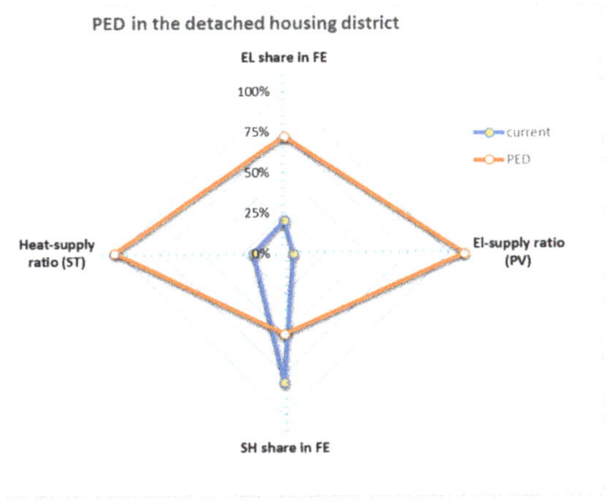


**Figure 5.** Key development indicators of transforming the detached housing district to PED (source: AIT). EL—electricity; SH—space heating; ST—solar thermal.

The exploitation of local renewables will evolve during the process of building refurbishment, which will proceed within a participatory process following a common agreement among the tenants/owners of the buildings in the considered district. With regards to the regulatory challenges for deploying PEDs, such a process can be triggered and accelerated by the applied incentives and promotion measures, beside the introduced regulations by the considered municipality, a governance challenge that needs to be tackled hand in hand with other challenges to enable the desired transition towards PED.

### 5.3. Flexibilization Need

The energy consumption of electricity and heat follows certain load profiles that change over days, weeks, and seasons showing periods of high demand (in the evening and during the cold winter days). In the current energy system infrastructure, the needed flexibility is offered completely by the supply system provided by the national/regional electricity grid (and gas and district heat grid), besides the big storage facility for fossil fuel. However, with the increased share of intermittent renewable energy sources (RES), additional flexibilisation options are neded, like local electric and heat storage. Since these measures have limited availability at the district scale (due to cost and operational management issues), the interaction with the regional energy supply infrastructure, e.g., regional electric grid and district heat grid, are indispensable in offering feasible solutions to help overcome the supply deficits, particularly in the winter time, and manage the energy surpluses produced in summer. Figure 6 presents the approach applied to handle flexibility needs to ensure adequate heat supply around the clock. Using typical normalised monthly load curves for space heating and hot water demands, beside the production curve of solar thermal in the considered district site, the figure demonstrates the periods of deficit and excess of heat supply through solar thermal energy. Based on the specified potential of ST, 76% of hot water and 24% of annual space heat demands can be covered by the installed solar thermal systems. However, cumulative heat excess and heat deficits that are compensated on an annual basis need further flexibility measures to manage the timely mismatch between demand and supply curves. The flexibility need is assumed to be achieved by interaction with the regional electric grid and district heating network, which are assumed to absorb heat excess in summer and provide compensation in winter. Other alternatives might be the availability of local heat storages. Similar behaviour is observed by comparing the PV supply curve with the electricity demand curve cumulating all electricity end-use (for HP, lighting, and appliances). In this case, flexibility will be ensured by interaction with the regional electric grid, beside locally available electricity storages.

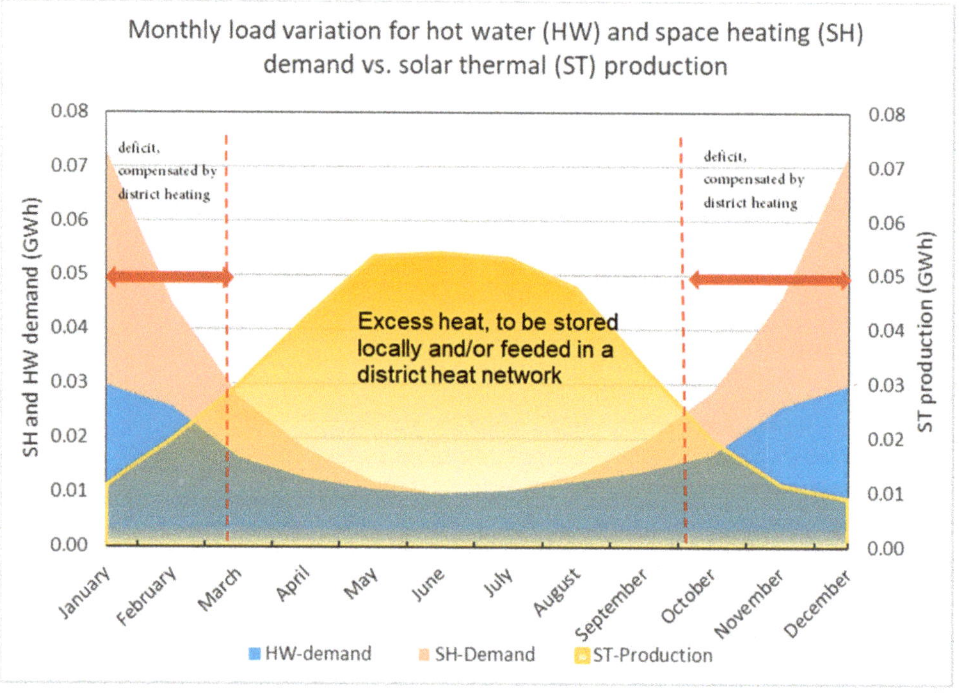


**Figure 6.** Monthly load curves for HW—hot water and SH—space heating demand and the supply curve of local ST—solar thermal: demonstrating the period of heat excess and deficit of the conceived PED in a detached housing district (source: AIT).

Table 3 summarizes comparatively the main results of evaluating the PED suitability of the four evaluated typologies in Vienna, elaborated in terms of select KPIs.

The results show that with the applied energy demand-supply measures, "Detached Housing" has the potential to become a PED with an annual positive energy balance of 110% of electricity supply with PV and 103% of heat supply with ST. The New Detached Housing fails to reach the PED-status by 10% due to the lower density of buildings, which means that fewer façades and roof surfaces are available for PV or solar thermal energy; the Medium Dense Housing fails by around 38%; and the Dense Inner-City by 55%. Hence, with further improvement in energy performance (there is still good potential for further energy performance improvement by enhancing the building shale insulation and using energy efficient window (e.g., adaptive thermochromic glazing with double glass). It is a matter of cost as such measures will go beyond the current standard of building refurbishment) and slightly increased use of open area for PV-panels, New Detached Housing can attain the status of PED. The results also reveal that the higher the buildings and the fewer open spaces in the district, the more difficult it becomes to achieve a PED status within the defined geographical boundary of the district. Figure 7 depicts the evolution of key indicators used to demonstrate the transition from the current state to a PED for the various considered typologies depicted for the specific final and useful energies.

**Table 3.** Key performance indicators (KPIs) for evaluating the energy demand-supply features of the selected typologies and their capability to become a PED.

| KPIs | Detached Housing 1961–1980 | | Dense Inner City before 1919 | | Medium Dense 1961–1980 | | Detached Housing 2006 | |
|---|---|---|---|---|---|---|---|---|
| | current | PED | current | PED | current | PED | current | PED |
| $FE/m^2$ | 179.4 | 60.7 | 183.0 | 73.4 | 199.6 | 76.6 | 130.9 | 55.5 |
| $UE/m^2$ | 165.0 | 102.4 | 168.8 | 106.8 | 166.4 | 104.3 | 121.4 | 90.7 |
| FE/Dw | 25,552.3 | 8640.4 | 11,912.4 | 4778.7 | 23,704.2 | 9095.8 | 19,504.0 | 8271.4 |
| UE/Dw | 23,497.8 | 14,582.6 | 10,990.3 | 6953.1 | 19,757.8 | 12,382.7 | 18,087.7 | 13,513.7 |
| $SH\text{-}UE/m^2$ | 128.8 | 70.1 | 112.4 | 57.6 | 125.5 | 68.0 | 85.7 | 58.8 |
| $SH\text{-}FE/m^2$ | 143.4 | 30.6 | 125.2 | 25.1 | 151.0 | 35.5 | 95.5 | 25.7 |
| EL share in FE | 20.4% | 72.3% | 24.6% | 67.6% | 22.2% | 70.8% | 23.9% | 71.5% |
| SH share in FE | 80.0% | 50.4% | 68.4% | 34.2% | 75.7% | 46.3% | 72.9% | 46.2% |
| $SR_{PV}$ | 5.2% | **109.7%** | 2.0% | **45.0%** | 3.0% | **61.6%** | 4.5% | **89.6%** |
| $SR_{ST}$ | 18.8% | **102.5%** | 12.2% | **60.6%** | 16.7% | **91.3%** | 20.4% | **97.0%** |

Dw—dwelling, EL—electricity, FE—final energy, UE—useful energy, SH—space heating, SR—supply ratio, calculated on an annual basis to cover electricity demands by local PV—Photovoltaics) and part of space heating and hot water by local ST—solar thermal.

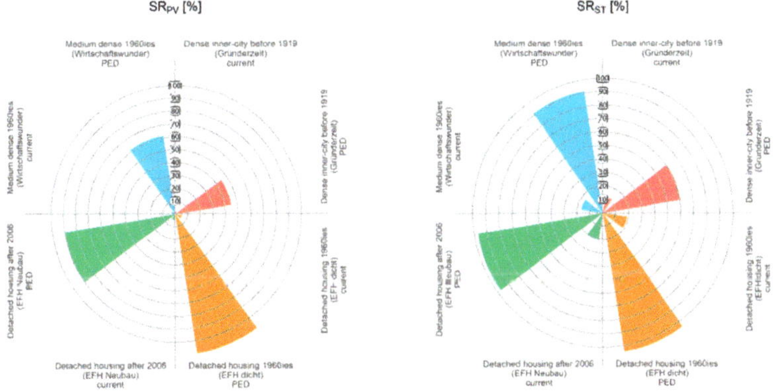


**Figure 7.** Supply ratio of PV—Photovoltaics and ST—solar thermal for the four analysed typologies in Vienna (source: AIT).

Finally, it is noteworthy that the realisation of a PED depends further on the cost of the proposed energy demand-supply measure of eco-refurbishment, electrification of the end-uses, and harnessing the locally available areas of rooftops, facades, and open places for PV and ST.

## 6. Conclusions and Outlook

The concept of PED represents a promising innovative solution with great leverage for the realization of urban energy and climate goals. It promotes urban change towards energy-optimized, integrative, and resilient cities. Given the fact that cities are responsible for about 75% of energy consumption and 80% of GHG emission—with the building sector alone responsible for around 28% [27,28]—the expected impact of implementing and deploying PEDs within Europe will have a significant role in enabling the targeted urban energy system transformation, given its high potential for energy-saving and local renewable energy integration.

Our study analysed the potential of four selected urban typologies to achieve a positive annual energy balance and thus become PEDs. Among the analysed typologies, the detached housing neighbourhood built between 1961 and 1980 shows the potential to become a PED under the given assumptions, and the more-energy efficient, but also significantly denser detached housing neighbourhood built after 2006 comes very close to the PED standard. The medium-dense neighbourhood built between 1960 and 1980 comes close to renewable self-sufficiency for heat supply, but is far off for electricity supply, while the inner-city neighbourhood built before 1919 does not show potential to become a PED. The results clearly show that PEDs require not only a very high level of energy efficiency, but also sufficient open spaces (including roofs and facades) for the local generation of renewable energy. If these two criteria are met, and the density of the neighbourhood does not become too high, the PED standard can be reached. An integrated planning process aligning urban and architectural design and energy planning is therefore key for the successful development of PEDs.

Considering that PEDs are expected further to generate an annual surplus of energy that can be used outside the district, PEDs will have a significant impact on the efficient and low-carbon urban energy system transformation once deployed among EU cities. In this regard, the substitution rate of buildings is low, compared to other energy-intensive sectors; thus, PEDs as a new innovative idea will help accelerate the transformation of building stock, given their high promising potential and the expected adaptation by building regulation and certification. This will directly result in accelerating the urban energy system, beside the expected indirect impact of the triggered technological innovation for improving the use of local renewables and flexibilization options, and their impact on the overall energy system.

The demonstrated screening and modelling approach is replicable in other areas and typologies due to its flexible bottom-up approach, which allows integrated energy demand-supply analysis at the district scale. However, there is the need to expand MAPED for improved modelling of flexibilization options, which need an integrated spatio-temporal assessment of the district energy system to capture existing synergies among the different use types (different dwelling types and building usages). It is therefore planned to include improved modelling of flexibilization, which will further enhance the applicability and replication potential of the established methodology. Future activities will deal with the question of designing and building local heat and power storages at the district scale. This effort requires additional considerations analysis based on techno-economic optimisation, considering the limits and constraints related to the interaction between the public heat and power grids and the applied regulatory framework. Further analysis effort is also needed to address the electricity demand for electric vehicle (EV) and the potential of the interaction of building and mobility as an additional flexibilization option.

**Author Contributions:** Conceptualization by H.-M.N., A.H., V.S; methodology, software, validation, and investigation—A.H., R.S., G.E., H.-M.N.; data curation—R.S., G.E., H.-M.N., A.H. writing—original draft preparation, A.H., H.-M.N., R.S., G.E.; writing—review and editing, H.-M.N., V.S., A.H.; visualization—R.S., G.E., A.H.; supervision—H.-M.N.; project administration—H.-M.N.; funding acquisition, H.-M.N. and G.E. All authors have read and agreed to the published version of the manuscript.

**Funding:** This journal article presents the results of the study "Aufbereitung städtischer Energiedaten für zukünftige Plus-Energiequartiere" (Preparation of urban energy data for future plus-energy districts), financed by the Republic of Austria via contract FN 252263a.

**Institutional Review Board Statement:** Not applicable.

**Informed Consent Statement:** Not applicable.

**Data Availability Statement:** Not applicable.

**Acknowledgments:** The authors would like to thank the Federal Ministry for Climate Action, Environment, Energy, Mobility, Innovation and Technology of Austria (BMK) and the Austrian Research Promotion Agency (FFG) for the excellent collaboration.

**Conflicts of Interest:** The authors declare no conflict of interest.

**Abbreviations**

| | |
|---|---|
| BESS | Battery energy storage system |
| BEVs | Battery-powered electric vehicles |
| BM | Biomass |
| BMK | Bundesministerium für Klimaschutz, Umwelt, Energie, Mobilität, Innovation und Technologie |
| DHCS | District heating/cooling system |
| DSM | Demand site management |
| CHP | Combined heat power |
| Dw | dwelling |
| EV | Electric Vehicle |
| FE | Final Energy |
| GHP | Geothermal heat pump |
| HDD | Heat Degree Days |
| HP | Heat pump |
| HW | Hot water |
| LHCs | Lighthouse cities |
| MAPED | Model for Analysis of Plus Energy District |
| NZEB | net-zero energy building |
| NZED | net zero energy district |
| NPC | Net present costs |
| PEDs | Positive Energy Districts |
| PE | Primary energy |
| PV | Photovoltaic |
| SC | Space cooling |
| SH | Space heating |
| SG | Smart grid |
| ST | Solar thermal |
| TRM | Technology Roadmap |
| UE | Useful energy |
| WH | Water heating |

# References


1. UN-SDGs. Sustainable Development Goals, 17 Goals to Transfer Our World. 2016. Available online: http://www.un.org/sustainabledevelopment/sustainable-development-goals/ (accessed on 23 April 2021).
2. European Commission. *SET Plan Action No. 3.2 Implementation Plan*; European Commission: Brussels, Belgium, 2018.
3. JPI Urban Europe. *PED Reference Framework*; JPI Urban Europe: Vienna, Austria, 2019.
4. European Parliament, Council of the European Union. *Directive 2010/31/EU of the European Parliament and of the Council of 19 May 2010 on the Energy Performance of Buildings*; Publications Office of the European Union: Luxembourg, 2010; Chapter 12; Volume 3, pp. 124–146. Available online: https://eur-lex.europa.eu/legal-content/EN/TXT/?uri=celex%3A32010L0031 (accessed on 23 April 2021).
5. European Commission. *SET Plan—Declaration of Intent on Strategic Targets in the Context of an Initiative for Smart Cities and Communities*; European Commission: Brussels, Belgium, 2016.
6. European Commission. *The European Green Deal*; European Commission: Brussels, Belgium, 2019.
7. Polly, B.; Kutscher, C.; Macumber, D.; Schott, M.; Pless, S.; Livingood, B.; Van Geet, O. *From Zero Energy Buildings to Zero Energy Districts*; National Renewable Energy Laboratory: Golden, CO, USA, 2016.
8. The Rocky Mountain Institute (RMI). *An Integrated Business Model for a Zero-Energy District*; RMI: Peshawar, Pakistan, 2017.
9. IEA EBC—Annex 83—Positive Energy Districts. 2021. Available online: https://annex83.iea-ebc.org (accessed on 23 April 2021).
10. Carlisle, N.; Van Geet, O.; Pless, S. Definition of a Zero Net Energy Community. Technical Report NREL/TP-7A2-46065. 2009. Available online: https://www.nrel.gov/docs/fy10osti/46065.pdf (accessed on 23 April 2021).
11. Polly, B. Zero Energy Districts. 2009. Available online: https://www.nrel.gov/docs/fy18osti/71424.pdf (accessed on 23 April 2021).

12. Hedman, Å.; Ur Rehman, H.; Gabaldón, A.; Bisello, A.; Albert-Seifried, V.; Zhang, X.; Guarino, F.; Grynning, S.; Eicker, U.; Neumann, H.-M.; et al. IEA EBC Annex83 Positive Energy Districts. *Buildings* **2021**, *11*, 130. [CrossRef]
13. Belussi, L.; Barozzi, B.; Bellazzi, A.; Danza, L.; Devitofrancesco, A.; Fanciulli, C.; Ghellere, M.; Guazzi, G.; Meroni, I.; Salamone, F.; et al. A review of performance of zero energy buildings and energy efficiency solutions. *J. Build. Eng.* **2019**, *25*. [CrossRef]
14. Everding, D.; Kloos, M. *Solarer Städtebau: Vom Pilotprojekt Zum Planerischen Leitbild*; Kohlhammer: Stuttgart, Germany, 2007.
15. Genske, D.; Jödecke, T.; Ruff, A. *Nutzung Städtischer Freiflächen für Erneuerbare Energien*; Erschienen: Bonn, Germany, 2009.
16. Everding, D.; Genske, D.; Ruff, A. Bausteine des energetisch-ökologischen Stadtumbaus. In *Energiestädte*; Springer Spektrum: Berlin/Heidelberg, Germany, 2020. [CrossRef]
17. Hegger, M.; Dettmar, J. *Energetische Stadtraumtypen: Strukturelle und Energetische Kennwerte von Stadträumen*; Fraunhofer IRB Verlag: Stuttgart, Germany, 2014.
18. Nitsch, D. *Wohngebietstypen*; Beiträge zur Stadtentwicklung: Vienna, Austria, 2016.
19. Stadt Wien. Katalog Wohngebietstypen 2016 Wien. Available online: https://www.data.gv.at/katalog/dataset/eab5ce18-827f-44a8-a206-e56f772dfa8b (accessed on 23 April 2021).
20. Statistics Austria. *Gebäude- und Wohnungsregister*; Statistics Austria: Vienna, Austria, 2016.
21. International Atomic Energy Agency (IAEA). *MAED, Model for Analysis of Energy Demand—User's Manual*; Computer Manual Series No. 18; IAEA: Vienna, Austria, 2006.
22. Statistics Austria, Overall Energy Consumption of Households, Vienna. 2019. Available online: https://www.statistik.at/web_en/statistics/EnergyEnvironmentInnovationMobility/energy_environment/energy/energy_consumption_of_households/index.html (accessed on 23 April 2021).
23. TABULA. WebTool. 2017. Available online: http://webtool.building-typology.eu/?c=all#bm (accessed on 23 April 2021).
24. Amtmann, M. *Reference Buildings, The Austrian Building Typology, Classification of the Austrian Residential Building Stock*; Scientific Report D 6.9; Austrian Energy Agency: Vienna, Austria, 2019.
25. Österreichisches Institut für Bautechnik (OIB). *Requirement for New Buildings and Major Renovations*; OIB RL 6 Energie-einsparung und Wärmeschutz; Österreichisches Institut für Bautechnik (OIB): Vienna, Austria, 2019.
26. Wiener Umweltanwaltschaft (WUA). *Zukünftige Chancen der Solarthermie in Wien*; WUA: Vienna, Austria, 2008.
27. International Energy Agency (IEA). *Tracking Buildings*; IEA: Paris, France, 2019.
28. Lucon, O.; Ürge-Vorsatz, D.; Zain Ahmed, A.; Akbari, H.; Bertoldi, P.; Cabeza, L.F.; Eyre, N.; Gadgil, A.; Harvey, L.D.D.; Jiang, Y.; et al. Climate Change 2014: Mitgation of Climate Change. In *Contribution of Working Group III to the Fifth Assessment Report of the Intergovern-Mental Panel on Climate Change*; Edenhofer, O., Pichs-Madruga, R., Sokona, Y., Farahani, E., Kadner, S., Seyboth, K., Adler, A., Baum, I., Brunner, S., Eickemeier, P., et al., Eds.; Cambridge University Press: Cambridge, UK; New York, NY, USA, 2014.

*Article*

# Positive Energy Districts and Energy Efficiency in Buildings: An Innovative Technical Communication Sheet to Facilitate Policy Officers' Understanding to Enable Technologies and Procedure

Tiziana Ferrante and Teresa Villani *

Department of Planning, Design, and Technology of Architecture, Sapienza University of Rome, 00196 Roma, Italy; tiziana.ferrante@uniroma1.it
* Correspondence: teresa.villani@uniroma1.it

**Abstract:** The Horizon 2020 framework programme is defining funding strategies for research and innovation projects in European cities and promoting policies and solutions for the transition to a competitive energy system at an urban scale. Given that Horizon Europe, thanks to the Driving Urban Transition Partnership, will fund RD&I projects regarding transitions to urban sustainability; how municipalities will implement different strategies is a relevant key to developing replicable models. We conducted this study on Italian cities through a mapping exercise on selected case studies. The aim was to provide a knowledge framework to municipalities undertaking sustainable urban development actions. We selected case studies based on energy efficiency in buildings, both in retrofits and new constructions. This highlighted how the adoption of multifaceted technological solutions blended well with each other, and led, not only, to satisfy the initial requirements, in terms of expected impacts from the single actions, but also provided relevant and replicable samples. For this, the analysis of solutions tested by different municipalities in the selected projects led to spreadsheets and indicators related to energy efficiency in buildings, which enabled a transition to a PED, which could facilitate an understanding of elements that must be clearly indicated in a preliminary design document (Directive 2014/24/UE).

**Keywords:** positive energy district (PED); enabling solution for PED transition; energy efficiency in buildings and real estate

**Citation:** Ferrante, T.; Villani, T. Positive Energy Districts and Energy Efficiency in Buildings: An Innovative Technical Communication Sheet to Facilitate Policy Officers' Understanding to Enable Technologies and Procedure. *Energies* **2021**, *14*, 8551. https://doi.org/10.3390/en14248551

Academic Editors: Paola Clerici Maestosi and Álvaro Gutiérrez

Received: 10 November 2021
Accepted: 10 December 2021
Published: 18 December 2021

**Publisher's Note:** MDPI stays neutral with regard to jurisdictional claims in published maps and institutional affiliations.

## 1. Introduction

### 1.1. Smart Cities and Positive Energy Districts: A European Commission Point of View

With the framework programme H2020, the European Commission, based on the Marseille and Toledo Declaration, council conclusions, opinions and the EU urban agenda [1–14], is defining funding strategies for cities addressing actions and programmes for sustainable urban development. Indeed, rapid population growth, deterioration of suburban areas and social inequalities, together with the increase in citizens' expectations of the quality of life and supplied services, make sustainable development policies a relevant key for energy saving and for the social participation of citizens. These topics have become the main focus in urban areas through promoting the transition to a competitive energy system based on several specific actions: reducing energy consumption and carbon footprints, supplying low-cost and low-carbon power, employing alternative fuels and mobile energy sources, employing a single and smart power network, researching new knowledge and technologies, sound decision-making, public commitment and an energy and ICT innovation market with capacity absorption [15,16]. Therefore, the attention on sustainable development drove, on one hand, to smart cities, and on the other, to PEDs (positive energy districts) [17,18].

Urban areas are indeed the main causes of climate change, and each action municipalities undertake for the future development of the city should contribute to characterize a positive global change. For this, municipalities play a key role in planning and decision-making for sustainable urban development [16].

It is clear that it was thanks to the cited documents that the European Commission developed an appropriate funding strategy to support sustainable development and sustainable urban areas. It is thanks to the contribution of a various set of stakeholders, such as the EERA Strategic Energy and Technology Plan (SET Plan) [19], JRC [20], IEA [21], JPI UE [22], the European Commission individuates, in the Horizon Europe framework programme that funding for urban sustainable development, with the Driving Urban Transition Partnership, has been directed towards positive energy districts, which represent one of the three pathways to facilitate urban transition [23].

The SET Plan, adopted by the European Union in 2008, was a first step to establish an energy technology policy for Europe; it was Europe's technology response to the challenges of meeting its targets on greenhouse gas emissions, renewable energy and energy efficiency. The integrated SET Plan identified 10 actions for research and innovation. The actions address the whole innovation chain, from research to market uptake, and tackled both financing and regulatory frameworks. Among the actions was Action 3.2, which stated, "Europe to become a global role model in integrated, innovative solutions for the planning, deployment, and replication of positive energy districts" [17]. The implementation plan, which was edited by smart cities and communities, focused on PEDs' requirements, such as an open innovation model for their planning, deployment and replication. In the TWG 3.2 implementation plan, cities were identified as the stakeholders who need to take a leading role in the integrated and holistic planning of PEDs, aligning it with their long-term urban strategies. Industries and organisations such as real estate development, construction companies, network operators, utility companies and many others, will play a vital role as solution providers.

Moreover, the Joint Programming Initiative Urban Europe—created in 2010 to address global urban challenges—thanks to the Strategic Research and Innovation Agenda 2.0 [24] and the White Paper "A Reference Framework for Positive Energy Districts and Neighbourhoods" [22], contributed to the definition of a positive energy district.

Additionally, a working group in the framework of the Joint Programming Initiative Urban Europe, analysed then collected the following in a booklet regarding PEDs [25]: PED projects in 61 urban areas, which were described according to key indicators in relation to PED projects (10 key indicators in the area of building/real estate) and to energy sustainability PED projects (5 key indicators and related sub-key indicators in the area of building/real estate). Subsequently, the above set of key indicators were extended in the ENEA research project [26] (Figure 1).

| PED Booklet — PED project (building/real estate) | | ENEA Key indicators — PED project (building/real estate) | |
|---|---|---|---|
| 1. City | | 1. City | textual description |
| 2. Project name | | 2. Project name | textual description |
| 3. Project status | 3.1 planned | 3. Project status | 3.1 planned |
| | 3.2 under construction | | 3.2 under construction |
| | 3.3 realized | | 3.3 realized |
| | 3.4 in operation | | 3.4 in operation |
| 4. Project start-end | | 4. Project start-end | textual description |
| 5. Contact | | 5. Contact | textual description |
| 6. Project website | | 6. Project website | textual description |
| 7. Size of project area | | 7. Size of project area | textual description |
| 8. Building structure | 8.1 newly built | 8. Building structure | 8.1 newly built |
| | 8.2 existing neighbourhood | | 8.2 existing neighbourhood |
| | 8.3 mixed | | 8.3 mixed |
| 9. Land use | | 9. Land use | textual description |
| 10. Financing | | 10. Financing | textual description |
| | | 11. Type of intervention | textual description |
| | | 12. procedure for implementation | textual description |
| | | 13. Ownership | textual description |
| | | 14. Competition notifying body (if public ownership) | textual description |
| | | 15. Financing type | textual description |
| | | 16. Financing amount | textual description |
| | | 17. Implementation phases | textual description |
| | | 18. Urban planning category | textual description |
| | | 19. Content | textual description |
| | | 20. Objectives | textual description |
| | | 21. Stakeholders | textual description |
| | | 22. Plan/Program Reference | textual description |
| | | 23. Involved Municipality sectors in public procurement | textual description |
| | | 24. Procedural requirements of special relevance | textual description |
| **Energy Sustainability PED project (building/real estate)** | | **Energy Sustainability PED project (building/real estate)** | |
| 25. Goals ambition | 25.1 positive energy | 25. Goals ambition | 25.1 positive energy |
| | 25.2 zero emission | | 25.2 zero emission |
| | 25.3 energy neutral | | 25.3 energy neutral |
| | 25.4 energy efficient | | 25.4 energy efficient |
| | 25.5 carbon free | | 25.5 carbon free |
| | 25.6 climate neutral | | 25.6 climate neutral |
| | 25.7 sustainable neighbourhood | | 25.7 sustainable neighbourhood |
| | 25.8 social aspects/affordability | | 25.8 social aspects/affordability |
| 26. Indicators/expected impact | | 26. Indicators/expected impact | |
| 27. Overall strategies of city/municipality connected with the project | | 27. Overall strategies of city/municipality connected with the project | |
| 28. Which factors have been included in implementation strategies? | 28.1 Local (renewable) resources | 28. Which factors have been included in implementation strategies? | 28.1 Local (renewable) resources |
| | 28.2 Regional energy system | | 28.2 Regional energy system |
| | 28.3 Mobility | | 28.3 Mobility |
| | 28.4 Buildings | | 28.4 Buildings |
| | 28.5 Materials | | 28.5 Materials |
| | 28.6 Refurbishment | | 28.6 Refurbishment |
| | 28.7 Sustainable production | | 28.7 Sustainable production |
| | 28.8 Sustainable consumption | | 28.8 Sustainable consumption |
| | 28.9 (Local) Governance | | 28.9 (Local) Governance |
| | 28.10 Legal framework | | 28.10 Legal framework |
| | 28.11 Business models | | 28.11 Business models |
| 29. Innovative stakeholder involvement strategies | | 29. Innovative stakeholder involvement strategies | textual description |
| | | 30. Type of energy supply | textual description |
| | | 31. Success factors | textual description |
| | | 32. Challenges/barriers | textual description |
| | | 33. Enabling technologies for the building | textual description |
| | | 34. Enabling technologies for the building's energy | textual description |
| | | 35. Indicators of expected impacts | textual description |
| | | 36. System solutions for energy production | textual description |
| | | 37. Energy flexibility solutions | textual description |
| | | 38. Production and supply of renewable energy | textual description |
| | | 39. Thermal energy storage | textual description |
| | | **Energy Efficency enabling solutions (building/real estate)** | |
| | | 40. Type of technological solutions | 40.1 Thermal coat |
| | | | 40.2 Plug & Play ventilated facade |
| | | | 40.3 Thermal break windows with triple glazing |
| | | | 40.4 Solar shields |
| | | | 40.5 Rainwater recovery system |
| | | | 40.6 Thermal activation |
| | | | 40.7 Renewable energy |
| | | | 40.8 Storage |
| | | | 40.9 Accumulation |
| | | | 40.10 Control and automation |
| | | | 40.11 ICT system energy monitoring |

**Figure 1.** Comparison of indicators and key indicators in the Booklet on PEDs and the ENEA research project.

### 1.2. A National Perspective on PED

At a national level the concept of PEDs are almost unknown among municipality public officers as there are no structured national events that discuss this topic, except the ones promoted by the national Italian delegate in the JPI Urban Europe to support alignment with the European dimension. Given that this was a national research activity coordinated by ENEA [26], which analysed in depth what types of data were needed to better explain the consistency of PED projects, based on these results we elaborated the innovative technical communication sheets.

Our manuscript presents the innovative technical communication sheets (selected examples from Milan, Florence and Trento), which were elaborated for 15 Italian case studies located in the seven selected municipalities, and presents them in a booklet of PEDs, they are also analysed by Bossi et al. [27]. While both cited documents refer to the European dimension with a set of a few data types, our manuscript presents the innovative technical communication sheets, which include more detailed data concerning technological solutions (project goals, expectations in terms of energy savings, energy class, initial/final energy rating, etc., which are individuated in Figure 2) and implementation processes (entity role in different phases, activities, instruments, etc., which are individuated in Figure 3), which refer to national experiences and facilitate national public officers' understanding of positive energy districts.

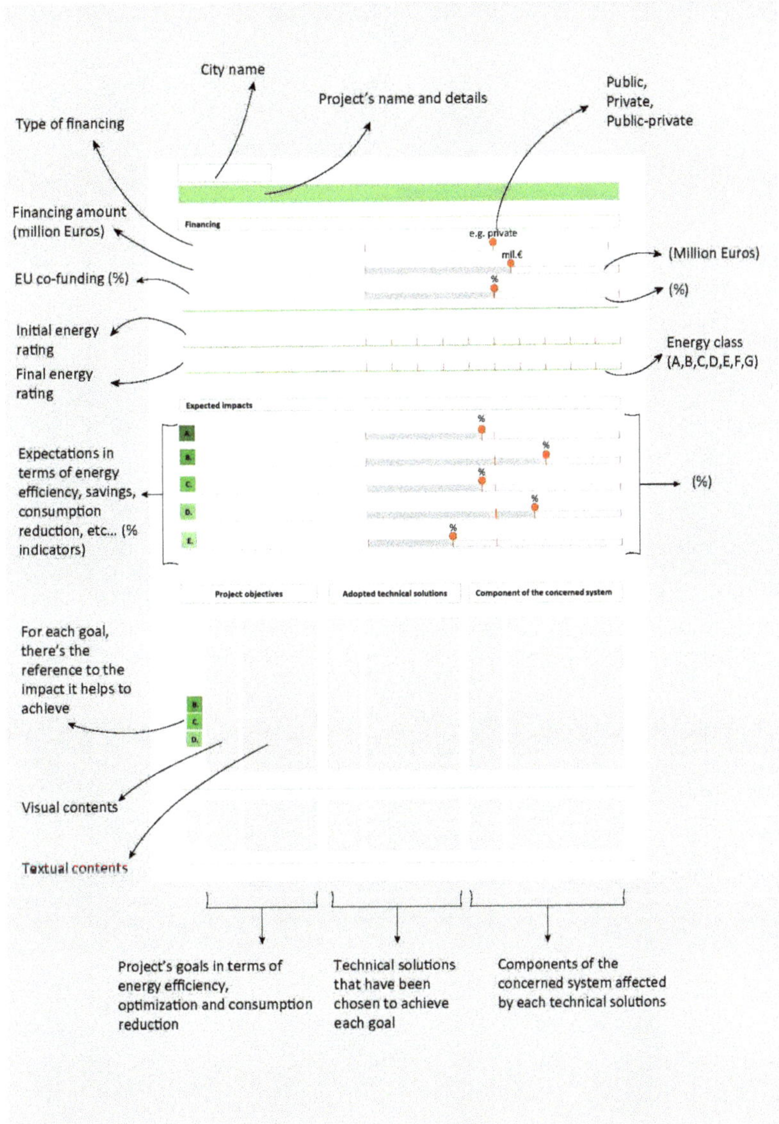



**Figure 2.** An example of a case-study project intervention sheet, the technological solutions.

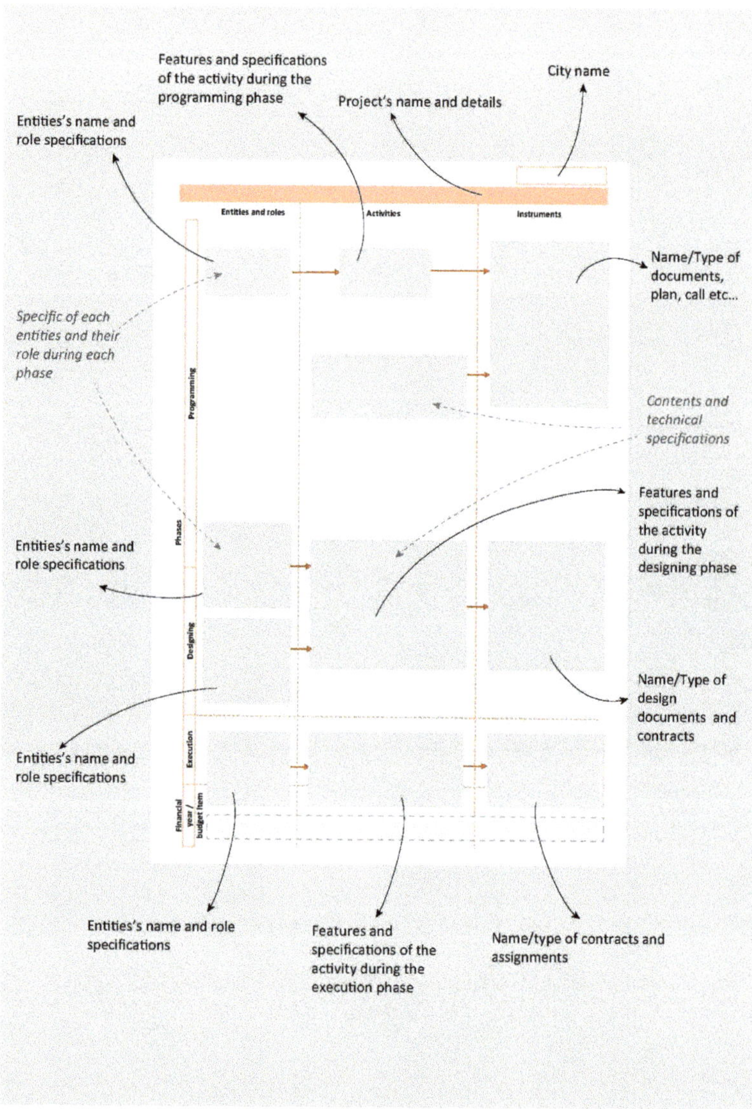



**Figure 3.** An example of a case-study project intervention sheet and the implementation processes.

### 1.3. Positive Energy District Functions

The guidance on PEDs includes 3 targets.

1. Efficiency: optimization of energy performance can reduce consumption in buildings and mobility infrastructures, including the existing building stock.
2. Flexibility: resilience of the regional energy system to carbon neutrality and 100% renewable energy.
3. Production: empowerment of relevant gas-emissions reductions.

### 1.4. Paper Content

Given this, and based on results of ENEA national research [26], the researchers verified that indicators used to describe PED experiences, at least at Italian level, were not developed enough, especially those related to energy efficiency in buildings. The results of the cited research highlights that we need more key indicators as well as new areas of investigations (6 areas of investigations and more than 100 key indicators) to individuate the innovative and integrated solutions in the planning/implementation phases, which contribute to successfully activate the transition towards PEDs. Our research focused on the area of energy efficiency in buildings/real estate, where the cited research individuated 35 new key indicators and 11 subindicators (Figure 1).

The novelty of our research activity relied on the creation of an innovative technical communication sheet to facilitate policy officers' understanding of enabling technologies and procedures that improve building energy efficiency in positive-energy-district projects. The manuscript presents the innovative technical communication sheets.

## 2. Materials and Methods

### 2.1. Materials

In our research we assumed, as a starting point, the following: (1) the contents of the Booklet on PEDs was collected and edited by the PED Programme Management of JPI Urban Europe and analysed 61 urban areas, 7 of them in Italy, namely Parma, Roma, Milano, Bolzano, Firenze, Lecce and Trento, proposing a characterization through a set of given elements; (2) results in the ENEA research implemented the original set in the Booklet on PEDs according to a more comprehensive approach, which considered each building as the result of a process characterized by phases, requirements and performance [28].

The starting point was to aggregate data that referred to a specific building, in order to obtain authorization to build, in a single technical communication sheet, then to implement the sheet with those characteristics to enable the building itself to be a physical node that enabled a positive energy district.

Our activity focused on the elaboration of an innovative technical communication sheet, which described, in an effective and easy way, key indicators related to buildings.

### 2.2. Methods

Research objectives: The research objectives aimed to individuate how to communicate the key indicators related to energy efficiency in PED building projects to policy officers in a municipality. These were based on the list of key indicators in the ENEA project, which were used to perform a deeper analysis on the selected case studies (Figure 1).

Research methodology: The research methodology was based on a theoretical perspective that highlighted technology-enabling factors for energy efficiency in buildings as nodes in a positive energy district, as well as rules, regulations and public procurement procedures for rehabilitation, recovery or new buildings.

The methodology was based on three phases.

1. Collection of objectives, aims and strategies in the PED building projects assumed as case studies;
2. Analysis of case studies according to the new key indicators related to energy efficiency in buildings/real estate;
3. Creation of technical communication sheets for easier understanding.

The paper presents the results of the research activities. This work focussed on identifying a technical communication sheet to facilitate the comprehension of key indicators that concern energy efficiency in buildings/real estate and the public procurement procedure. This work also aimed to facilitate the comprehension and understanding by civil servants in municipalities, which aimed to activate the transition to positive energy districts.

According to each case study, and to the above-mentioned key indicators, it was necessary to present analysis results on energy-efficiency solutions in buildings as well as on public procurement procedures in each case study. This helped facilitate the comparison and identification of enabling factors for the transition to a PED, as related to the specific topic.

Thus, two types of technical communication sheets were created.

To define the spreadsheet structure, the study considered the data collected in the 7 Italian municipalities, which were already selected as "in transition to a PED" in the Booklet on PEDs [25], and 15 energy-efficiency interventions on buildings (7 already indicated in the Booklet of PEDs and 8 new ones), which were identified as case studies specifically referring to energy efficiency in buildings/real estate projects.

Then according to the type of indicators, we developed one technical communication sheet describing energy-efficiency solutions in building/real estate case studies and a second one describing the public procurement process.

The contents of the technical communication sheet describing energy-efficiency solutions in the buildings and real estate case studies included the following:

- General information such as the city's name and the project's name and details;
- Quantitative and qualitative data, such as type of financing, financing amount, EU co-funding (% indicators);
- Data on energy class upgrades, such as initial energy rating and final energy rating (energy class from A to G);
- Information on expected impacts, such as expectations in terms of energy efficiency, savings, consumption reduction, etc. (% indicators);
- Data on adopted technology solutions in response to project objectives (project goals in terms of energy efficiency, optimization and consumption reduction and their impact when achieved) (Figure 2).

The contents of the technical communication sheet describing the public procurement process included the following:

- General information such as the city's name and the project's name and details;
- Building phases (programming, designing, execution);
- Financial information.

For each phase we identified entities and roles (which were specific for each entity and role during each phase), activities (in terms of procedures, rules, requirements, financing, etc., contents and technical specifications, instruments (the name/type of documents, plans, calls, etc., and the name/type of design documents), contracts and assignments) (Figure 3).

The proposed technical communication sheet was undoubtedly an effective way to present results and to compare building solutions and public procurement procedures.

The frames were designed with the aim of facilitating the use of the content information by the principles of information visualization. This improved the cognitive process for an understanding of the spreadsheet's content through a balanced use of visual and textual codes, which were also used to achieve a different type of communication effectiveness, depending on the nature of the information (qualitative and quantitative) and on the recipients.

Specifically, for frames of the implementation process of operations, a lot of the collected data in the spreadsheets concerned descriptive topics. These needed a textual language for communication, but, at the same time, must communicate an organized sequence of phases concerning operators and proper tools for their specific activities.

In this case, to represent the complexity of the building process and the public procurement (with its relevant dataset) of the analysed case studies, we chose a representation that integrated text-based information organized within flow diagrams that facilitated a logical sequence with links among areas.

The aim of the proposed graphic organisation was to disseminate this information to municipal officials/civil servants (competent departments and offices) who wanted to start an urban development plan for driving ecological transitions and PEDs. By learning from this information, which also displayed a more immediate visual representation, officials could take out useful instruments for the selection of policy and decision making on funding strategies, appropriate procurement processes and key documents to ensure the quality of the operations, in terms of energy efficiency and sustainability, and to identify any issues (diagram bugs) and anticipate solutions.

In this way the exploratory analysis of the data benefitted from the visual representation through fast information communication that otherwise hid within the spreadsheets. This positively affected the target in replicating the case studies' procedures. Visual attributes, such as colour, size, proximity and visual representation of quantitative data, as well as textual content, were used for the frames on technological solutions, to convey complex data that would otherwise require huge cognitive processing. We collected the set of data through the desk activity on innovative solutions for the energy efficiency of the case studies and on expected impacts. We then reorganized them into a consistent form that was able to communicate the target context, taking well into account the key recipients as municipal technical office chiefs, architects, etc.

## 3. Results

The method for the frame implementation was to report a first phase of all information taken from a critical reading of the analysed projects and to gradually remove redundant information, or information liable to further technical insight. This way allowed us to bring out only first-level data, which were useful to identify, with an immediate representation, the impact of the positive solution within the PEDs. (Figure 4). In addition, this helped technicians consider the content's consistence for their replicability.

We carried out work on the spreadsheet affordance, i.e., those real or perceived properties which were self explanatory, thus simply showing them to informed recipients supported its multiple uses in terms of data interaction.

The use of visual representation for quantitative context data made it easier to use the preconditions for starting energy-efficiency operations based on the need for funding. In particular it showed the impacts of EU grants.

The different levels of colour saturation and intensity facilitated the process of visual recognition for most relevant data.

The choice of the use of icons made effective mental representations, facilitated fast communication and, in any case, kept a close link between visual and textual terminology.

This information, together with the experience of recipients, should be able to produce knowledge on the project objectives, on the specificities of adopted solutions and on the key components of the building system for energy efficiency. Knowledge was therefore the main objective of the communication process. This empowered technicians to express a meaningful consideration on data and to develop the technical knowledge for the replication of the solutions.

In order to assess the on-field effectiveness of the frames we were holding, we had many online meetings with specific stakeholders of the involved municipalities, from which we received useful feedback on how to improve the communicativeness of the frames; by implementing content and graphics, for example.

Indeed, these frames were also prototypes on which to test the accessibility of information content and its actual use by checking compliance with replicable solutions. The obtained feedback, and the application of information visualization principles, were also crucial for the ongoing design of the digital database.

Moreover, the communication outline of the frames was particularly effective for a comparative assessment between the procedures of the operations in more cities and the technological solutions for energy-efficiency operations in the case studies.

Indeed, the objective comparison between homogeneous datasets was visually facilitated by reducing the error during data comparison, which were carried out on spreadsheets.

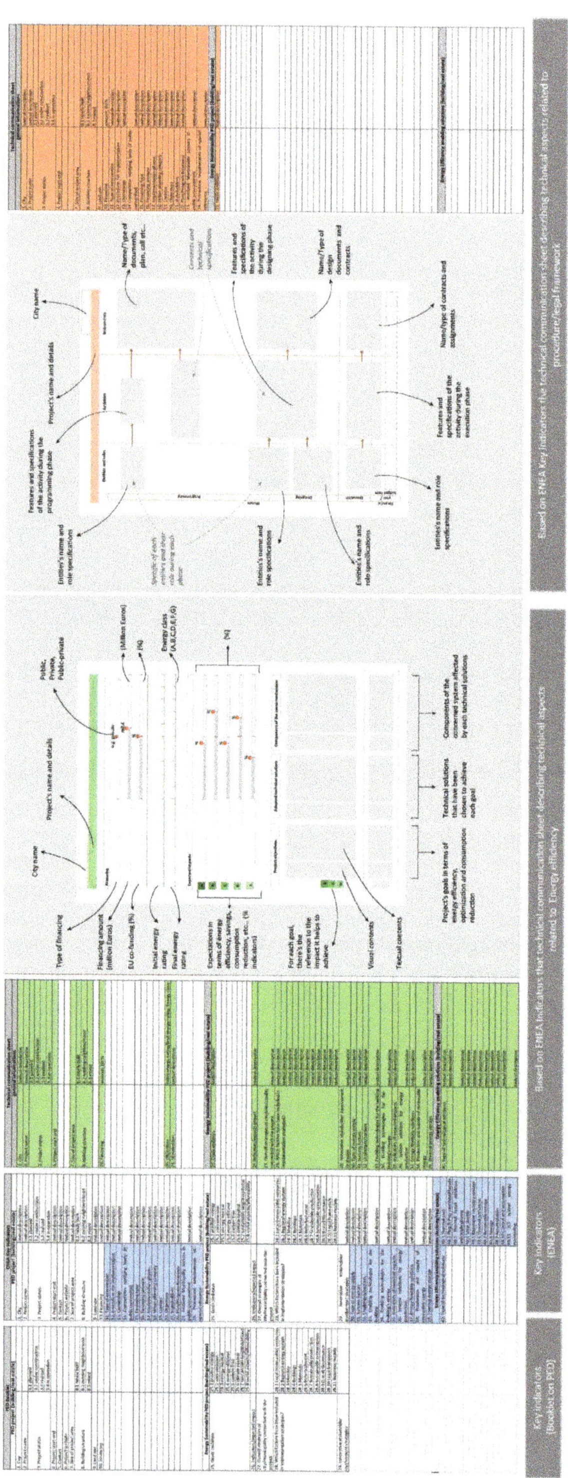

**Figure 4.** Comparison of a technical data sheet and a technical communication sheet.

For the frame of technological solutions, the selected communication outline could highlight the "weight" of the results achieved by the different solutions, compared within each indicator, related to the expected impacts. Such impacts needed a close link, not only to adopt solutions, but also for the building system components that were involved in the surveyed energy-efficiency operations (Figure 5).

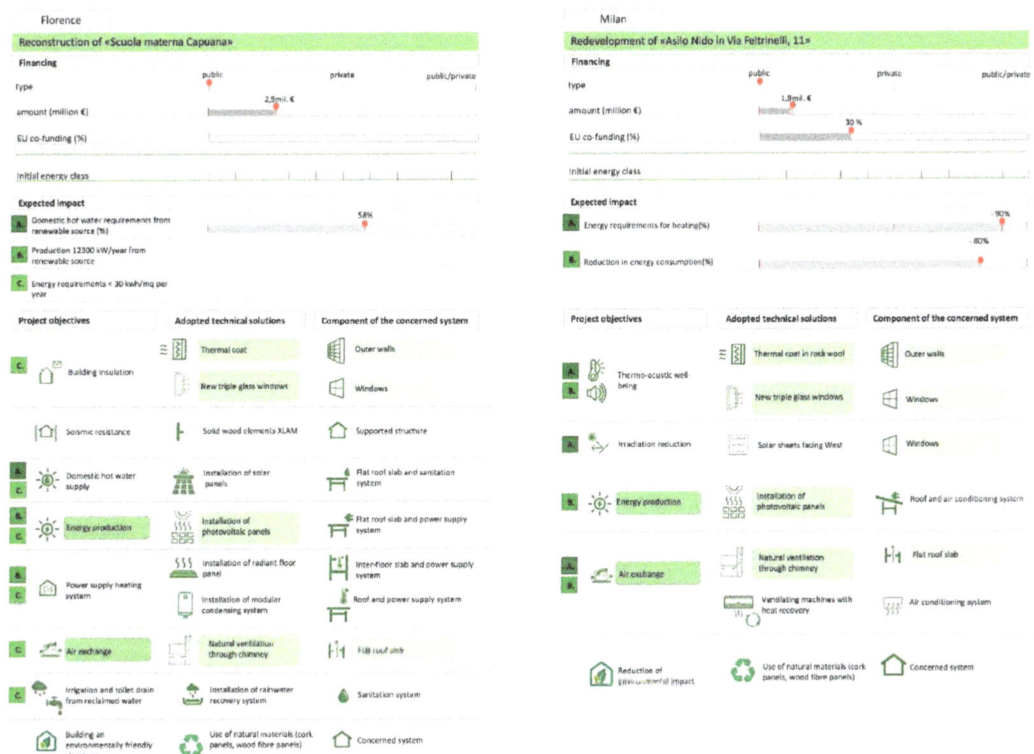

**Figure 5.** Milan and Florence case study: comparative analysis of technological solutions.

For the frame of the execution process through to the comparison of flow diagrams, we could focus on the critical phases of the building process and on the potential for different instruments the municipalities chose for each activity, as well as the extensive impact on the quality and achievement of the PED target (Figure 6).

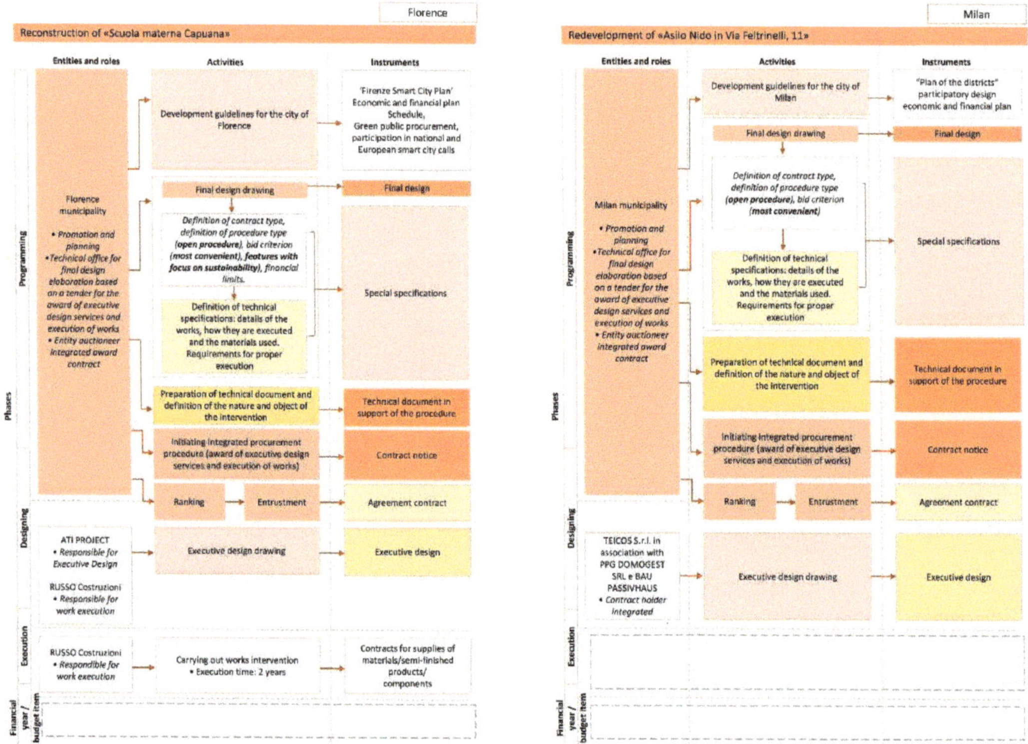

**Figure 6.** Milan and Florence case study: comparative analysis of the implementation processes.

### 3.1. Technical Communication Sheets on Technological Solution

Systematization of the information collected in the spreadsheets allowed a univocal interpretation of the different analysed projects, in terms of the innovative technological solutions and the processes that empowered planning, development, implementation and management of different actions. The importance of the analysis of technological solutions concerned the topic of energy efficiency: the design of such actions was the result of strategic policies and the use of specific solutions to achieve established objectives. Such solutions took place to achieve a sustainable architecture based on a low environmental impact of the building through resource saving and pollution reduction in all lifecycle phases. There was also a positive approach to "on-site" energy production using specific systems that used renewable sources. Within the study carried out in this first phase, we found that projects stood out for the adoption of different solutions that were able to improve energy performance; for this, we identified an in-depth study on how adopted technological choices interacted with the whole building system. We reported the sheet sample of the case study of the city of Trento (Figure 7).

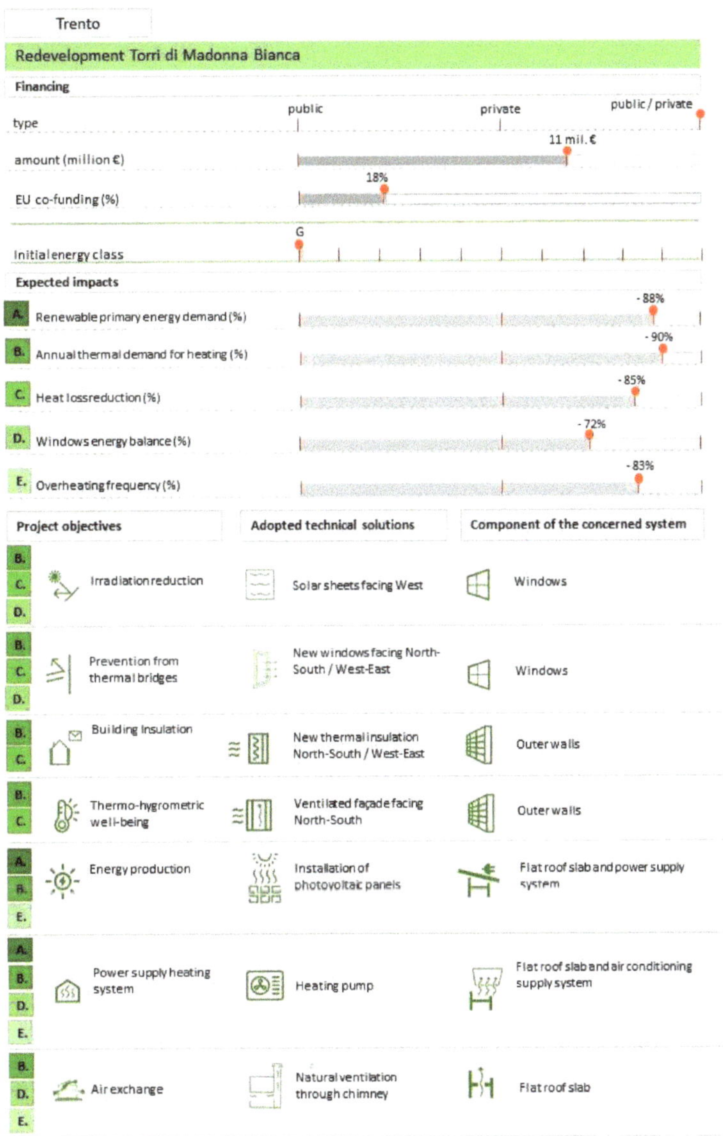



**Figure 7.** Trento case study project intervention: the technological solutions.

We conducted the project's survey and linked it closely with the analysis of each building process.

*3.2. Technical Communication Sheets on Implementation Processes*

In this section we discuss the spreadsheet on implementation processes concerning building rehabilitation. In reference to the procedures we outlined, with the involvement of municipalities and stakeholders, we identified specific actions that cities developed. The project variety depended on factors including the programming methods of actions, the financing policies and the stakeholders' involvement through participatory processes.

We reported an example of the described sheet in the case study of the city of Trento (Figure 8).

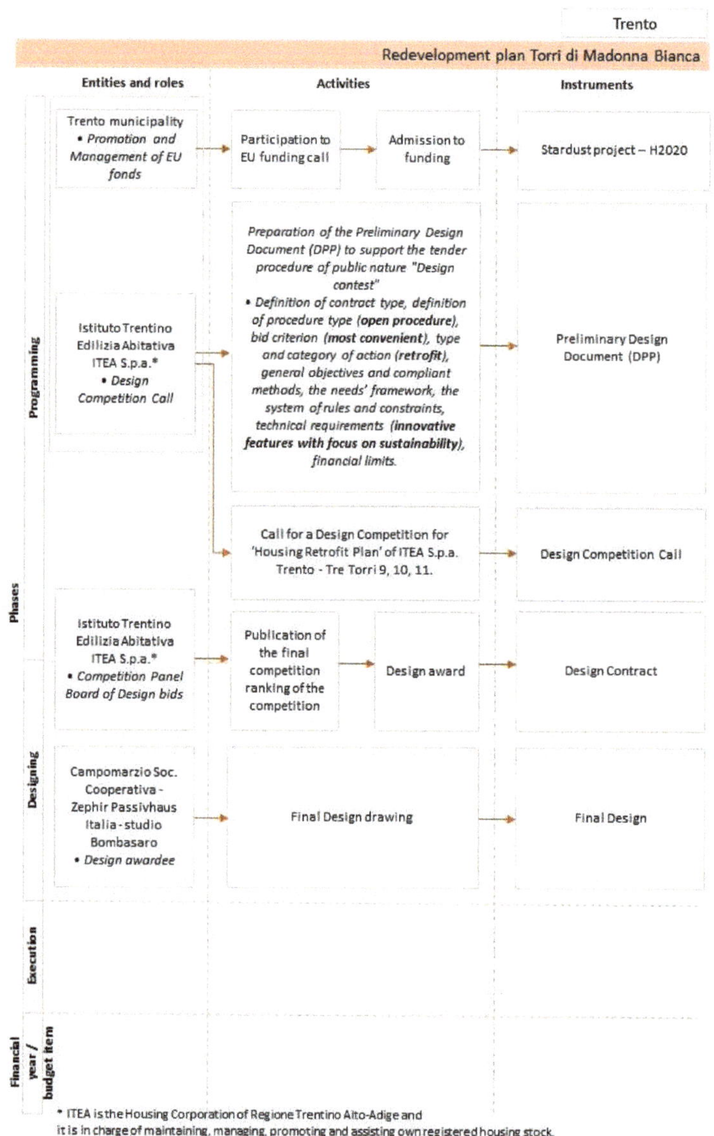


**Figure 8.** Trento case study project intervention: the implementation processes.

### 3.3. New Set of Indicators Comparing the Requirement in the Preliminary Design Document with the Technological Solutions Adopted

The result obtained for each project, from these two sheets, was not only useful for specific content but especially for the link we could detect between the two observation areas. Taking Trento, again, as an example case study, the analysis of the implementation processes led us to highlight the relevance of the documents used since the programming phase. Indeed, the quality of the pre-design document, which we found complete and clear in the compliance to specific requirements of energy sustainability, highlights several indicators, which proved to be valuable during the preparatory phase prior to design (Figure 9); these indicators did not limit the building scale, but detected all factors of the action.

| Technological solutions - Trento – Torri di Madonna Bianca | | |
|---|---|---|
| **Requirement** | **Objective** | **Adopted technological solutions** |
| Integration with landscape/environment surroundings | | Gres panels preserving materic and colour features of the original towers |
| Shorter construction site duration | | Use of dry and precast assembly systems |
| Compliance to housing needs | Residents staying home during operations onsite | |
| Thermal insulation of matt surfaces | Prevention from thermal bridges | Thermal loss reduction |
| | Energy efficiency (energy class upgrade) | Continuous insulating jacket of building shell – Optimized passive solar supply |
| Execution flexibility | Fast execution phases | Plug&Play ventilated facade system installed out of the floors |
| | Integrability of elevation components | New panels shall be with the same size of the current ones |
| Fire safety | Use of non-combustible materials | Specific materials (mineral wool, gres, metal, glass) |
| New transparent partitions | New windows | New steel windows |
| Usability/accessibility | Easy opening/closing of doors/windows | Natural ventilation system through automatized opening |
| Shielding systems installation | Proper natural lighting | Independent solar sheets for ensuring full protection from solar irradiation |
| System integration | Preserving original features | Additional storey for the new system |
| New air-conditioning systems | Energy efficiency upgrade | Photovoltaic system from 70kWpeak<br><br>Heating pump |
| Preserving original features | Check of outwards facing and surroundings | Terrace with panoramic view on the valley |
| Regulated architectural composition flexibility | Limited works on some features | |
| Flexibility of adopted technical solutions | | |
| Improving thermal comfort | Reduction of thermal bridges of balconies | Use of high-performance insulating materials |
| Complying with size rules | Adding cubature | "A" Energy class upgrade allows a cubature bonus of 350 mc for each housing block |
| Privacy and safety | Safety of outer partition components | Use of proper railings |
| | Ensuring privacy among balconies | Use of partitions |
| Enhanced accessibility and usability | Architectural barriers overcome – recognizability of main entrances | Use of high quality materials for accessible entrance |
| Easy maintenability of green areas | Maintaining through time common and private green areas | |
| Highest re-use of building waste (CAM) | | Check of use of building materials |
| Cost-effectiveness during and after construction | | Cost-effectiveness for using a standard (photovoltaic) system |

**Figure 9.** Indicators described in the preliminary design document for the Trento case study.

The design phase, which followed the public procurement procedure, was tendered for by the design contract, in accordance with the DPP requirements. This produced an easier selection criteria procedure for the bid and for the selection of especially energy-efficient technological solutions. Strategic policies were also a driver to meet the requirements as reported among expected impacts in terms of energy. Thus, the overall process quality that included sustainability topics from the early stages of promotion and programming to all implementation tools, should clearly result in success of the action with a sensible design of the technological solutions in compliance with the PED criteria.

## 4. Discussion

Although our research activity will support further in-depth investigation, the proposed frame that describes PED projects, appears promising for the organization of information and has a clear focus on strategies and solutions, which will facilitate public officers in understanding the main characteristics of PED projects.

Several discussions are ongoing among most prominent European networks (EERA JPSC PED modules, SET Plan Action 3.2 PED Programme/DUT PED pillar, COST Action PED-EU-NET, IEA EBC Annex 83, UERA PED WG, PED-related SCC01 projects, H2020 SCC01 TG Replication, SCALE, Smart Cities Marketplace) to define a common and shared definition of a PED, as well as to individuate key indicators that capture the true PED essence.

No matter what the key indicators will be, our research activity highlighted the importance of effective communication.

With increasing numbers of city authorities embracing the PED concept, and an increasing number of PED-related projects, there is a lack in communication characteristics, results, aims and goals in PED projects, in knowing that clear and effective communication facilitates comparison, evaluation and replication.

Our research activity was aligned with these contents and our aim was to fulfil the existing gap by contributing, with our results, methods of organizing information to facilitate understanding of PED projects.

## 5. Conclusions

Although our research will further develop an in-depth analysis of other aspects of cities from our case studies (e.g., identification and qualification of urban ecosystem stakeholders and sectors of competence with their involvements at national, regional, provincial and local levels), the results presented in this paper seem interesting and in line with ongoing European debate.

There is a need for effective communication on PED case studies that highlight information related to strategies and solutions implemented by the municipalities to facilitate the transition to PEDs, in the specific field of energy efficiency in the building/real estate sectors.

Effective communication on findings related to PED pathways will encourage and facilitate synergies among urban ecosystem stakeholders by activating virtuous communication processes and improving the understanding of actions to support PED transition.

Indeed, the new set of indicators identified by the ENEA optimizes the understanding of the technicalities of PED projects, thanks to a national research project we carried out to identify an effective way to communicate results and findings in an appropriate way to public officers.

Among technicalities, the most prominent was within the building process (planning of sustainable actions, design development, implementation and management), which represented the fundamental activity public officers take to move from ideas to reality.

That is why our main aim was to communicate the effectiveness of solutions and procedures that could be assumed as a set of replicable good practices among public officers involved in technical offices or sectors within municipalities: from several tender, technical and administrative documents, as well as financing budgets required for activating public tenders.

Finally, a particular attention was paid to the planning phase of the building process and to the contents of the preliminary design document where public officers within municipalities expressed requirements and addressed the choices in a meaningful way.

Indeed, a clear and effective qualification of the demand arose from the necessary conditions for the implementation of technological and financial solutions. In fact, the key role of public officers of municipalities was clear from the coherence and specificity of the technical planning documents.

Without their involvement and support there was no chance in the area of positive energy districts, due to the complexity of the involved areas of expertise in road mapping urban transition strategies.

**Author Contributions:** Conceptualization, T.F. and T.V.; methodology, T.F.; validation, T.F. and T.V.; formal analysis, T.V.; investigation, T.F. and T.V.; resources, T.F. and T.V.; data curation, T.V.; writing—original draft preparation, T.F. and T.V.; supervision, T.F.; project administration, T.F. All authors have read and agreed to the published version of the manuscript.

**Funding:** Project funded by Ricerca di Sistema, Report RdS/PTR (2020)/040.

**Institutional Review Board Statement:** Not applicable.

**Informed Consent Statement:** Not applicable.

**Data Availability Statement:** The data presented in this study are available on request from the corresponding author.

**Acknowledgments:** We would like to thank the public officials of the municipalities included in the Booklet on PEDs and the Architect Eleonora Di Manno, who participated in data collection, analysis and report.

**Conflicts of Interest:** The authors declare no conflict of interest.

## References


1. 'Marseille Declaration' Adopted at the Informal Ministerial Meeting of Ministers Responsible for Urban Development on 25 November 2008. Available online: http://www.eib.org/attachments/jessica_marseille_statement_en.pdf (accessed on 9 December 2021).
2. Toledo Declaration, Adopted at the Informal Ministerial Meeting of Ministers on Urban Development of 22 June Adopted 2010 in Toledo. Available online: http://www.ccre.org/docs/2010_06_04_toledo_declaration_final.pdf (accessed on 9 December 2021).
3. Territorial Agenda of the EU 2020, Agreed at the Informal Ministerial Meeting of Ministers responsible for Spatial Planning and Territorial Development of 19 May 2011 in Gödöllő. Available online: http://ec.europa.eu/regional_policy/en/information/publications/communications/2011/territorial-agenda-of-the-european-union-2020 (accessed on 9 December 2021).
4. Urban Agenda for the EU—'Pact of Amsterdam', Agreed at the Informal Meeting of EU Ministers Responsible for Urban Matters on 30 May 2016 in Amsterdam. Available online: http://ec.europa.eu/regional_policy/sources/policy/themes/urban-development/agenda/pact-of-amsterdam.pdf (accessed on 9 December 2021).
5. The General Affair Council of the European Union—Council Conclusions on the Urban Agenda for the EU, 24 June 2016, the Council of the European Union. Available online: http://www.consilium.europa.eu/en/press/press-releases/2016/06/24/conclusions-eu-urban-agenda/ (accessed on 9 December 2021).
6. Council Conclusions on the Objectives and Priorities of the EU and Its Member States, Adopted for the Third United Nations Conference on Housing and Sustainable Urban Development (Habitat III), 2016. Available online: http://www.consilium.europa.eu/en/press/press-releases/2016/05/12/conclusions-on-habitat-iii/ (accessed on 9 December 2021).
7. Report from the Commission to the Council on the Urban Agenda for the EU, COM (2017) 657 final. 2017. Available online: https://ec.europa.eu/regional_policy/sources/policy/themes/urban/report_urban_agenda2017_en.pdf (accessed on 9 December 2021).
8. Opinion of the Committee of the Regions 'Towards an Integrated Urban Agenda for the EU', Adopted on 25 and 26 June 2014. Available online: http://eur-lex.europa.eu/legal-content/EN/TXT/PDF/?uri=CELEX:52013IR6902&from=PL (accessed on 9 December 2021).
9. Opinion of the Committee of the Regions 'Concrete Steps for Implementing the EU Urban Agenda', Adopted on 7 April 2016. Available online: http://eur-lex.europa.eu/legal-content/EN/TXT/?uri=CELEX%3A52015IR5511 (accessed on 9 December 2021).
10. Committee of the Regions—Commission for Territorial Cohesion Policy and the EU Budget 'The Follow-Up Strategy on the Implementation of the Urban Agenda for the EU'. 2016. Available online: http://memportal.cor.europa.eu/Handlers/ViewDoc.ashx?doc=COR-2016-04284-00-01-TCD-TRA-EN.docx (accessed on 9 December 2021).

11. European Economic and Social Committee Opinion: The 2030 Agenda—A European Union Committed to Sustainable Development Globally, Adopted on 20 October 2016. Available online: http://www.eesc.europa.eu/en/our-work/opinions-information-reports/opinions/2030-agenda-european-union-committed-support-sustainable-development-goals-globally-own-initiative-opinion (accessed on 9 December 2021).

12. Resolution of the European Parliament of 3 July 2018 on the Role of Cities in the Institutional Framework of the Union (2017/2037(INI)). Available online: http://eur-lex.europa.eu/legal-content/EN/TXT/PDF/?uri=CELEX:52018IP0273&rid=2 (accessed on 9 December 2021).

13. Report of the Parliamentary Committee on Regional Development on the Urban Dimension of EU policies (2014/2213(INI)), Adopted on 17 June 2015. Available online: https://www.europarl.europa.eu/doceo/document/A-8-2015-0218_EN.html (accessed on 9 December 2021).

14. Opinion of the European Committee of the Regions—'Implementation Assessment of the Urban Agenda for the EU', Adopted on 5 July 2018. Available online: http://memportal.cor.europa.eu/Handlers/ViewDoc.ashx?pdf=true&doc=COR-2017-06120-00-00-AC-TRA-EN.docx (accessed on 9 December 2021).

15. Østergaard, P.A.; Maestosi, P.C. Tools, technologies and systems integration for the Smart and Sustainable Cities to come. *Int. J. Sustain. Energy Plan. Manag.* **2019**, *24*, 1–6. [CrossRef]

16. European Union. Regional Policy. City of Tomorrow. Challenges, Visions, Ways Forward. 2011. Available online: https://ec.europa.eu/regional_policy/sources/docgener/studies/pdf/citiesoftomorrow/citiesoftomorrow_final.pdf (accessed on 2 November 2021).

17. SET-Plan Action 3.2 on Smart Cities and Communities. Available online: https://jpi-urbaneurope.eu/wp-content/uploads/2018/09/setplan_smartcities_implementationplan.pdf (accessed on 2 November 2021).

18. Joss, S.; de Jong, M.; Schraven, D.; Zhan, P. Sustainable–Smart–Resilient–Low Carbon–Eco–Knowledge cities; making sense of a multitude of concepts promoting sustainable urbanization. *J. Clean. Prod.* **2015**, *109*, 25–38. [CrossRef]

19. The EU SET-Plan. EERA. Available online: https://www.eera-set.eu/eera-in-the-eu/set-plan.html (accessed on 2 November 2021).

20. Shnapp, S.; Paci, D.; Bertoldi, P. *Enabling Positive Energy Districts Across Europe: Energy Efficiency Couples Renewable Energy, EUR 30325 EN*; Publications Office of the European Union: Luxembourg, 2020. [CrossRef]

21. IEA EBC—Annex 81—Data-Driven Smart Buildings. Available online: https://annex81.iea-ebc.org/ (accessed on 2 November 2021).

22. Hinterberger, R.; Gollner, C.; Noll, M.; Meyer, S.; Schwarz, H.-G. White Paper on PED Reference Framework for Positive Energy Districts and Neighbourhoods, JPI Urban Europe and SET-Plan 3.2 Programme on Positive Energy Districts. 2020. Available online: https://jpi-urbaneurope.eu/wp-content/uploads/2020/04/White-Paper-PED-Framework-Definition-2020323-final.pdf (accessed on 2 November 2021).

23. Proposal for the European Partnership Driving Urban Transitions. 2020. Available online: https://ec.europa.eu/info/sites/default/files/research_and_innovation/funding/documents/ec_rtd_he-partnerships-driving-urban-transitions.pdf (accessed on 2 November 2021).

24. Joint Programming Initiative Urban Europe. Strategic Research and Innovation Agenda 2.0. 2019. Available online: https://jpi-urbaneurope.eu/wp-content/uploads/2019/02/SRIA2.0.pdf (accessed on 2 November 2021).

25. Joint Programming Initiative Urban Europe. Europe Towards Positive Energy Districts. 2020. Available online: https://jpi-urbaneurope.eu/wp-content/uploads/2020/06/PED-Booklet-Update-Feb-2020_2.pdf (accessed on 2 November 2021).

26. Ferrante, T.; Villani, T. Il Ruolo delle Municipalità nei Progetti di Ricerca, Sviluppo e Innovazione: Transizione Verso Aree Urbane Sostenibili, ENEA, Ricerca di Sistema Elettrico, Report RdS/PTR (2020)/040. 2020. Available online: https://www.enea.it/it/Ricerca_sviluppo/lenergia/ricerca-di-sistema-elettrico/accordo-di-programma-MiSE-ENEA-2019-2021/tecnologie/tecnologie-per-la-penetrazione-efficiente-del-vettore-elettrico-negli-usi-finali/local-energy-district (accessed on 2 November 2021).

27. Bossi, S.; Gollner, C.; Theierling, S. Towards 100 Positive Energy Districts in Europe: Preliminary Data Analysis of 61 European Cases. *Energies* **2020**, *13*, 6083. [CrossRef]

28. Koskela, L.; Huovila, P.; Leinonen, J. Design management in building construction: From theory to practice. *J. Constr. Res.* **2020**, *3*, 1–16. [CrossRef]

*energies*

*Article*

# An Evaluation Framework for Sustainable Plus Energy Neighbourhoods: Moving Beyond the Traditional Building Energy Assessment

Jaume Salom [1,*], Meril Tamm [1], Inger Andresen [2], Davide Cali [3], Ábel Magyari [4], Viktor Bukovszki [4], Rebeka Balázs [4], Paraskevi Vivian Dorizas [5], Zsolt Toth [5], Clara Mafé [6], Caroline Cheng [7], András Reith [4,8], Paolo Civiero [1], Jordi Pascual [1] and Niki Gaitani [2]

1. Thermal Energy and Building Performance Group, Catalonia Institute for Energy Research (IREC), 08930 Sant Adrià de Besós, Spain; mtamm@irec.cat (M.T.); pciviero@irec.cat (P.C.); jpascual@irec.cat (J.P.)
2. Department of Architecture and Technology, Norwegian University of Science and Technology (NTNU), 7491 Trondheim, Norway; inger.andresen@ntnu.no (I.A.); niki.gaitani@ntnu.no (N.G.)
3. Department of Applied Mathematics and Computer Science, Technical University of Denmark (DTU), 2800 Kongens Lyngby, Denmark; dcal@dtu.dk
4. Advanced Building and Urban Design Ltd., 1123 Budapest, Hungary; magyari.abel@abud.hu (Á.M.); bukovszki.viktor@abud.hu (V.B.); balazs.rebeka@abud.hu (R.B.); reith.andras@abud.hu (A.R.)
5. BPIE—Buildings Performance Institute Europe, 1040 Brussels, Belgium; vivian.dorizas@bpie.eu (P.V.D.); zsolt.toth@bpie.eu (Z.T.)
6. Housing Europe, 1000 Brussels, Belgium; clara.mafe@housingeurope.eu
7. SINTEF Community, 7034 Trondheim, Norway; caroline.cheng@sintef.no
8. Research Group 'Well Being Research Incubator', University of Pécs, 7624 Pécs, Hungary
* Correspondence: jsalom@irec.cat

**Citation:** Salom, J.; Tamm, M.; Andresen, I.; Cali, D.; Magyari, Á.; Bukovszki, V.; Balázs, R.; Dorizas, P.V.; Toth, Z.; Mafé, C.; et al. An Evaluation Framework for Sustainable Plus Energy Neighbourhoods: Moving Beyond the Traditional Building Energy Assessment. *Energies* 2021, *14*, 4314. https://doi.org/10.3390/en14144314

Academic Editor: Patrick Phelan

Received: 7 May 2021
Accepted: 13 July 2021
Published: 17 July 2021

**Publisher's Note:** MDPI stays neutral with regard to jurisdictional claims in published maps and institutional affiliations.

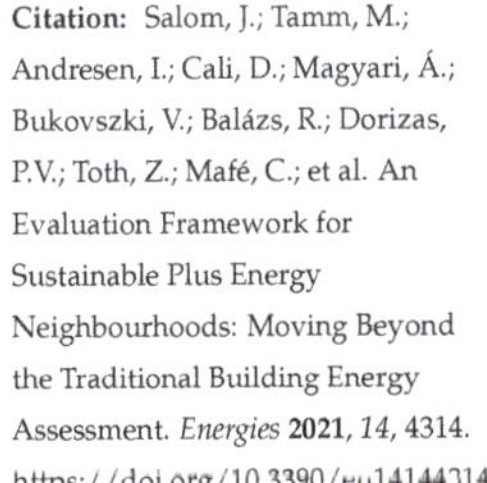

**Abstract:** There are international activities and on-going initiatives, particularly at the European level, to define what Positive Energy Districts should be, as the driving concept for the urban transition to a sustainable future. The first objective of the paper is to contribute to the on-going and lively debate about the definition of the notion of Sustainable Plus Energy Neighbourhood (SPEN), which highlights the multiple dimensions when talking about sustainability in districts moving beyond the traditional and strict building energy assessment. Based on a holistic methodology which ensures the consideration of the multidimensional nature and goals of SPEN, the paper outlines an evaluation framework. The evaluation framework defines the key performance indicators distributed in five categories that consider energy and power performance, GHG emissions, indoor environmental quality, smartness, flexibility, life cycle costs and social sustainability. This framework is designed to be implemented during integrated design processes aiming to select design options for a neighbourhood as well within during the operational phase for monitoring its performance. Further work will include the implementation and validation of the framework in four real-life positive energy neighbourhoods in different climate zones of Europe as part of syn.ikia H2020 project.

**Keywords:** neighbourhoods; positive energy districts; sustainable urban areas; energy production; energy efficiency; energy flexibility; economic costs; indoor environmental quality; social performance

## 1. Introduction

Over the course of syn.ikia H2020 project [1], four real-life Sustainable Plus Energy Neighbourhoods (SPEN) tailored to four different climatic zones will be developed, analysed, optimised and monitored, demonstrating the functionality of the plus-energy neighbourhood concept in Europe. When it comes to the implementation of sustainable development in the construction sector, the focus has started shifting from individual buildings (micro-scale) to districts and cities (meso- and macro-scale) (Figure 1). The idea of shifting scales is based on believing that the sustainability challenge has to do with more than just buildings; it includes interrelationships between buildings, open spaces, users, infrastructures and transport networks [2].

The transition from single buildings to neighbourhoods brings a need to fully understand, assess and regulate the potential for energy flexibility including clusters of buildings at an aggregated level. A cluster of buildings implies that several buildings can be located physically next to each other, or digitally connected having one common instance (usually named the aggregator) controlling and managing their energy flexibility. Aggregation of the energy flexibility from several buildings is required to ensure a significant impact to the energy systems and grids, in contrast to the limited energy flexibility of a single building, e.g., in Net ZEBs [3].

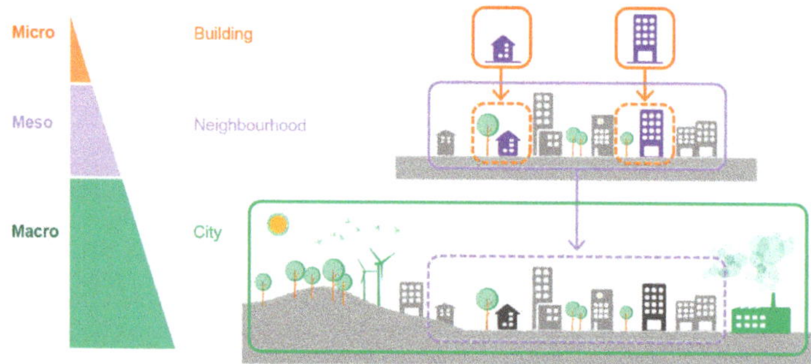


**Figure 1.** Identification of the neighbourhood scale. Source: IREC (2021), adapted from [4].

The development of sustainable plus energy neighbourhoods is aligned with the broad concept of Positive Energy District (PED) stated by the implementation plan of SET Plan Action TWG 3.2, thus anticipating the concept of PEDs highlighted in the European Partnership Driving Urban Transition to a sustainable future [5]. It is inspired by discussions within the European Innovation Partnership on Smart Cities and Communities (EIP-SCC) supported by the European Commission, and especially by the initiative on Positive Energy Blocks and the "Zero Energy/Emission Districts" mentioned in the TWG 3.2 declaration of intent [6]. In this context, a PED is considered as an "energy-efficient and energy-flexible urban area or groups of connected buildings which produce net zero greenhouse gas emissions and actively manage an annual local or regional surplus production of renewable energy". A PED requires integration of different systems, interaction between buildings and users, and other mobility, ICT and energy systems. A PED should secure the energy supply and wellbeing considering social, economic and environmental sustainability aspects. This concept is the result of several working groups and on-going initiatives at European level. JPI Urban Europe [6] conducts the programme "Positive Energy Districts and Neighbourhoods for Sustainable Urban Development" which aims to support the planning, deployment and replication of 100 Positive Energy Neighbourhoods by 2025 as one of the means to face the global urban challenges of today. The European Energy Research Alliance Joint Program on Smart Cities [7] help to define the European research landscape on smart cities based on the experience from H2020 smart city lighthouse projects and other national actions. The European Innovation Partnership on Smart Cities and Communities is merged with the "Smart Cities Information System" (SCIS) in one single platform named the "Smart Cities Marketplace" with the aim to bring cities, industries, SMEs, investors, researchers and other smart city actors together [8]. The European Regions Research and Innovation Network has specific working groups on smart cities and energy and climate change [9]. Eurocities [10] coordinates activities based on the knowledge sharing among more than 190 cities in 38 counties in Europe. The International Energy Agency EBC Annex 83 [11] started to work on developing an in-depth definition of PED and researching on technologies and planning tools for the decision-making process related to positive energy districts. Finally, the COST Action CA19126 "Positive Energy Districts European Network" [12] will establish a PED innovation eco-system to facilitate open sharing of knowledge and support the capacity building of new generation PED professionals, early Career Investigators as well as experienced practitioners. Outside Europe, NREL developed a guide to show how the implementation of district-scale high-performance scalable strategies can be suitable approaches to achieve deep energy savings, to increase affordability, reduce emissions and improve resilience [13].

Despite the above-mentioned efforts, there is still no standard definition of the PED concept, even if a shared definition developed by the EERA JPSC together with the JPI Urban Europe, integrates a wide vision of different projects and programs in Europe. It gathers the main characteristics of the PED projects and precursors of PEDs [14]. According to this work, up to four categories of PEDs have been established based on two main aspects: the boundaries and limits of the PED in order to reach a net positive yearly energy balance and the energy exchanges (import/export) in order to compensate for energy surpluses and shortages between the buildings and the external grid [15]. The four categories are:

1. Auto-PED (PED autonomous): "plus-autarkic", net positive yearly energy balance within the geographical boundaries of the PED and internal energy balance at any moment in time (no imports from the hinterland) or even helping to balance the wider grid outside;
2. Dynamic-PED (PED dynamic): net positive yearly energy balance within the geographical boundaries of the PED but dynamic exchanges through the boundary compensate for momentary surpluses and shortages;
3. Virtual-PED (PED virtual): net positive yearly energy balance within virtual boundaries of the PED and also dynamic exchanges with outside to compensate surpluses and shortages; and
4. Candidate-PED (pre-PED): no net positive yearly energy balance within the geographical boundaries of the PED but energy difference is provided by the market with certified green energy.

All of the described categories of PEDs are based on the accomplishment of a yearly positive energy balance, measured in greenhouse gas emissions, with use of renewables within the defined boundaries. Auto-and Dynamic-PEDs are the only categories where a net positive energy balance is achieved and Candidate-PED should compensate the energy difference with imported certified energy from outside the boundary. The difference between Auto-PED and Dynamic-PED is that the first does not need to import energy at any time. The difference between Dynamic-PED and Virtual-PED is that the latter defines the boundaries of the PED as virtual and they are not limited to a geographical area.

The ISO 52000-1:2017 is the overarching EPB (Energy Performance of Buildings) standard, providing the general framework of the EPB assessment based on primary energy as the main indicator. In order to evaluate the Positive Energy Balance, the set of EPB standards play a key role to assess the energy performance as defined in the recast of the Energy Performance of Buildings Directive (EPBD) (The EPBD and the EED have been amended by Directive (EU) 2018/844, which entered into force on 9 July 2018). Each of the five EPB standards describes an important step in the assessment of the energy performance of single buildings and a building portfolio [16] From the amended (2018) text of EPBD Annex 1, point 1: "Member States shall describe their national calculation methodology following the national annexes of the overarching standards, namely ISO 52000-1, 52003-1, 52010-1, 52016-1, and 52018-1, developed under mandate M/480 given to the European Committee for Standardisation (CEN)". When the positive energy balance assessment moves from a single building to a group of buildings (Building Portfolio) at the neighbourhood scale, new considerations are needed in terms of integrating urban and energy planning to evaluate the overall energy performance. Furthermore, neighbourhoods include other technological, spatial, regulatory, financial, legal, environmental, social and economic perspectives, but also barriers and challenges, which are not fully covered nor planned by the EPBD and EPB standards. Among them, barriers for changing household behavioural need to be understood and analysed to design measures that make households to contribute in mitigating climate change [17].

Moving from the building to the neighbourhood scale fits very well with the smartness imperative of exploiting all of the potential from collaborative approaches. The SPEN concept includes a profound integration and interoperability between buildings, urban spaces [18], the grid and infrastructures, but also with their governance. For instance, when focusing on a set of buildings, it is required to consider a common technical system whose energy performance considers the aggregated performance. Aggregation articulates synergies and discloses higher potential for smart and mutual interaction [19]. Therefore, the neighbourhood scale will foster sustainability through economies of scale, aggregation synergies (e.g., the deployment of flexibility and integration) and a considerable involvement of stakeholders and communities. The vision of future buildings described in [20] sees buildings as active components of larger districts which should be able to adapt to changing environmental conditions and occupancy, supporting well-being and using resources efficiently. The authors of [20] proposed a framework with 14 metrics to drive the transformation of the building stock with 100-year targets. The authors of [21] stresses the need for clear, comprehensible and structured definitions, including KPIs, after reviewing 144 scientific publications and analysed 35 terminologies on zero emission neighbourhoods, positive energy districts and similar concepts of climate friendly neighbourhoods.

The objective of this paper is to present and define the concept of Sustainable Plus Energy Neighbourhood (SPEN) which highlights the need of considering mutual interaction between the built environment, the inhabitants and the nature (Figure 2). The definition aims to contribute to the on-going debate for a common vision of what a PED should consider. Through a multidimensional analysis to address complexity in neighbourhoods, this paper outlines an assessment framework for the performance evaluation of SPEN. The selection of the main assessed categories and Key Performance Indicators (KPIs) have been based on a holistic and comprehensive methodology which highlights the multiple dimensions of sustainability in the built environment. The contents of the paper are based on the work developed in the syn.ikia project [22] with extended details on the methodology applied, revised definitions and concise and synthetic presentation of the metrics. Section 2 presents the definition of the SPEN and Section 3 describes the methodology applied to select the different categories and indicators ensuring multidimensionality. The core key performance indicators are presented for each dimension in the results section, followed by the conclusions of the paper.

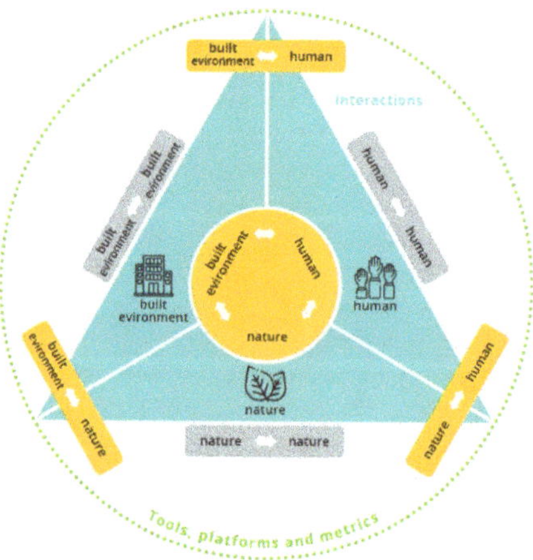


**Figure 2.** Interactions between humans, nature and the built environment Source: Reith, A. [23].

## 2. Definition of Sustainable Plus Energy Neighbourhoods (SPEN) and System Boundaries

### 2.1. Definition of SPEN

The syn.ikia definition of a SPEN follows a similar basis for Positive Energy Buildings (PEB), but the geographical boundary is physically or digitally expanded to the entire site of the neighbourhood, including local storage and energy supply units (Figure 3). Users, buildings and technical systems are all connected via a Digital Cloud Hub (HUB) and/or common energy infrastructures. The SPEN framework includes a strong focus on cost efficiency, indoor environmental quality, spatial qualities, sustainable behaviour, occupant satisfaction, social factors (co-use, shared services and infrastructure and community engagement), power performance (peak shaving, flexibility and self-consumption) and greenhouse gas emissions.

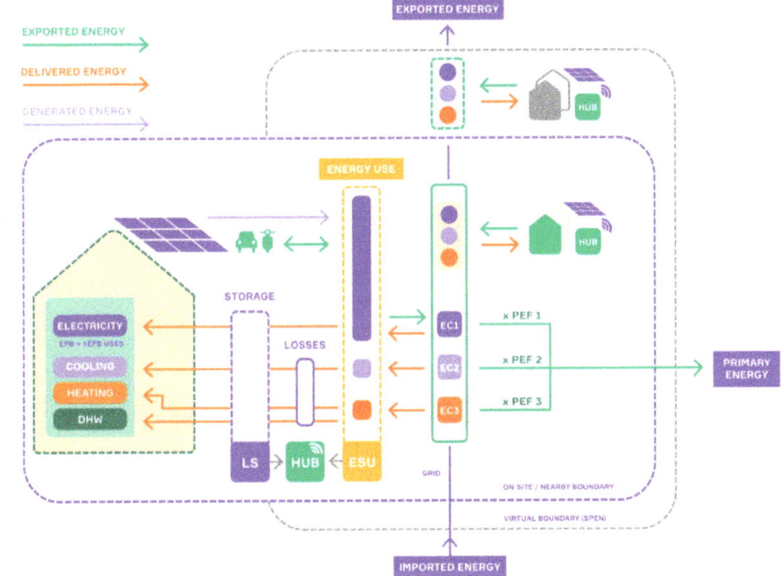


**Figure 3.** SPEN schematic as defined within syn.ikia project. Source: Skogen, syn.ikia project [4].

SPEN is defined as a group of interconnected buildings (The neighbourhood concept in syn.ikia project refers to, but is not limited to the Building Portfolio definition within the ISO52000 that considers a set of buildings and common technical building systems whose energy performance is determined by considering their mutual interactions [SOURCE: ISO 52000-1:2017, 3.1.6]) with associated infrastructure (infrastructure includes grids and technologies for exchange, generation and storage of electricity and heat. Infrastructure may also include grids and technologies for water, sewage, waste, mobility, ICT, and Energy Management System (EMS)), located within a limited geographical area and/or a virtual boundary. A SPEN aims to reduce its primary energy use towards zero over a year and an increased use and generation of renewable energy. A SPEN—a highly energy efficient neighbourhood with a surplus of energy from renewable sources—should focus on the following key-points:

- A SPEN is embedded in an urban or regional energy system and is driven by renewable energy to provide energy security and flexibility of supply;
- A SPEN is based on well-designed and high-efficient energy measures aiming to reduce the local energy consumption below the amount of locally produced renewable energy;
- A SPEN enables increased use of renewable energy by offering optimised flexibility and by managing consumption and storage capacities responding to demand;

- A SPEN couples the built environment with sustainable energy generation, use, and mobility (e.g., EV charging) to create added value for the citizens;
- A SPEN optimally use advanced technologies and materials, local RES, and other solutions as local storage, smart energy management systems, demand-response, user interaction ICT, etc.; and
- A SPEN offers affordable housing, high-quality indoor environment, and well-being for the human beings.

SPEN should be focused in five areas named the 5D areas:

- **Decentralisation:** neighbourhoods, as flexibility providers, allow higher penetration of renewable energy sources into the grid and increase flexibility;
- **Democracy:** empowered and conscious users having access to affordable homes and high-quality neighbourhoods;
- **Decarbonisation:** climate neutral, meaning efficient districts with a minimal final energy consumption and generating a surplus of energy from renewable sources;
- **Digitalisation:** ICT based neighbourhoods integrating smart networks that provide well-managed built environment for the citizens; and
- **Design:** highly attractive energy-efficient urban neighbourhoods by means of an integrated energy, architectural and outdoor spaces design that increase their market uptake.

5S strategies are identified as the ones that facilitate the achievement of SPENs:

- **Save:** reducing the neighbourhood net energy consumption by using solutions based on a total life cycle cost analysis;
- **Shave:** facilitating peak shaving through load shifting, control, and storage, thus reducing the size of energy supply installations, increasing self-consumption of renewable energy, and reducing the stress on the grid;
- **Share:** sharing of resources such as energy, infrastructure, and common spaces with neighbours;
- **Shine:** ensuring high quality architecture, creating good indoor and outdoor environments and solutions that make the occupants and the community proud of their neighbourhood;
- **Scale:** benefitting from large-scale effects of the neighbourhood scale to replicate the solutions.

*2.2. SPEN and Different Level of System Boundaries*

There is a continuous discussion of where to define the system boundaries, i.e., what energy elements to include in the balance when developing and defining PEDs. From a technical point of view, a SPEN is characterised by achieving a positive energy balance within a given system of boundaries according to an Energy Community scheme [24]. There are multiple ways to cover the RES generation in a SPEN. Moving from the single building boundary to the neighbourhood scale widens the on-site generation possibilities significantly. The scale is not restricted to on-site boundaries and, when using the SPEN smartness attributes, a SPEN may expand beyond the physical boundaries of the community. The SPEN boundary may address two different levels:

- **Functional Boundary:** On one hand, a functional boundary addresses the spatial-physical limits of the building portfolio and the neighbourhood. On the other hand, it addresses the limits with regards to the energy grids considering them as a functional entity of the neighbourhood that they serve. (e.g., a district heating system that can be considered as a functional part of the neighbourhood even if its service area is substantially larger than the heating sector of the building portfolio in question). Renewable share of the energy infrastructures (e.g., electricity from the grid) is included in the balance with the use of appropriate conversion factors from final energy to primary energy or $CO_2$ emissions.

- **Virtual Boundary:** This addresses the limits in contractual terms, e.g., including a renewable energy generation system owned by the occupants but situated outside the geographical boundaries (e.g., an offshore wind turbine owned through shares by the community).

According to these boundaries descriptions, and aligned to the draft definition of PEDs from the EERA JPSC working group and JP Urban Europe [15], the net positive yearly energy balance of a SPEN will be assessed within the functional or virtual boundaries. Thus, a SPEN will achieve a positive yearly energy balance having dynamic exchanges within the functional/virtual boundaries, but in addition, it will provide a connection between buildings inside the boundaries of the neighbourhood. In a SPEN, buildings can be digitally connected by means of a digital cloud hub (HUB), sharing ICT infrastructure and energy management systems.

### 3. Methodology: Ensuring Multidimensionality in Selection of Indicators

Assessment of SPENs can be a challenging exercise, since neighbourhoods, energy systems on a neighbourhood scale and sustainability itself are complex to evaluate [25,26]. Thus, evaluation frameworks that are already present agree on the fact that neighbourhoods can only be evaluated when taking the combined effect of multiple factors into account. From a completely different point of view, it is really important for different legislative measures and interventions to consider the actual problems, and react to the existing practice. Otherwise it is possible, that their effectiveness will be limited. Consequently, a holistic approach need to be used to ensure the consideration of appropriate measures and mitigation of known obstacles [27]. Decision-making processes consists of numerous independent factors differing by stakeholders and other levels, i.e., political interests, personal beliefs, market orientation, etc. [28]. In a proper decision making process, all of these drivers need to be considered. Therefore, it is needed to design a holistic, multidimensional assessment framework considering all elements of a SPEN that can diagnose and adapt to numerous district resources, cater for different users and market conditions and initiate commercial arrangements between partners in and out of the SPEN. Creating connected and equitable targets across multiple dimensions is also described as the " energy trilemma" by the World Energy Council (WEC). It is described as a combination of three equally important factors: environmental sustainability, energy security and energy equity. Valdes [25] mentioned that it is critical to review the robustness of the indicators. Ensuring multidimensionality on the selection level can be done with the help of four design consideration: avoidance of selection bias (Diversity analysis); avoidance of anchoring bias (Multiple valid impact chains) [26]; avoidance of overreliance on available and measurable data (Multiple valid impact chains) [28]; and avoidance of multicollinearity (D-separation) [29,30].

To make sure, that all the previously described considerations are ensured, directed acyclic graphs (DAGs) are built, and the KPIs are tagged along different aspects. Through a diversity analysis, tagged KPIs are able to ensure the heterogeneity and avoidance of selection bias, while with the help of a method called "d-separation" and by creating at least two impact chains for each goal, DAGs can ensure the rest of the design considerations. Tagging key performance indicators (KPIs) is a widely used approach and it is able to help ensuring the heterogeneity and spread of KPIs across different aspects. Usually, sustainability focused indicator development frameworks are based on the 'three pillars of sustainability': environmental, social and economic pillars [31]. Furthermore, there are other aspects when we consider the intersections of the previously mentioned pillars: livable, equitable and the viable dimensions as presented in Figure 4.

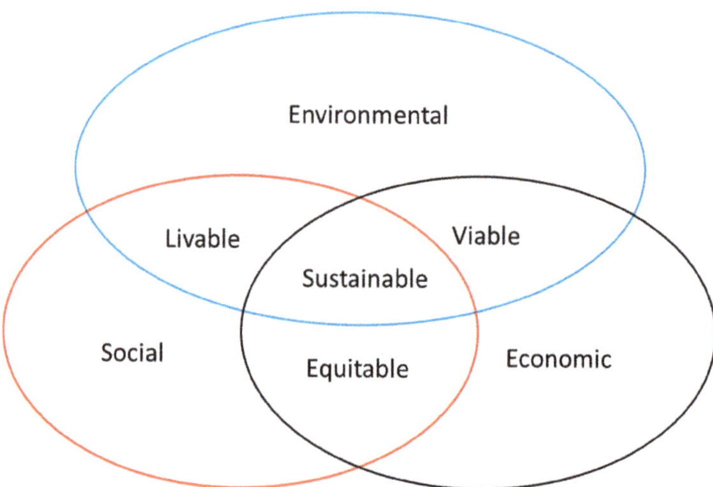


**Figure 4.** Pillars of sustainability and their intersections by Tanquay et al. [31].

KPIs however do not always fit these six categories, and forcing them into these categories may cause misunderstandings. Furthermore the boundaries between these categories are not completely defined. It is therefore more appropriate to categorize the KPIs along different, more SPEN definition relevant aspects. These are defined as followings:

- Domain of sustainability (Social, Economic, Environmental);
- Life cycle stage (Design, Operation);
- Scale (Household, Building, Neighbourhood);
- Functionality (Core, Sub);
- Type (Categorical, Numerical, Boolean, Index);
- Authority (Occupant, Facility manager, Grid operator, Policy developer, Building owner);
- Relation to the five main SPEN focus areas defined as the 5D: Design, Decentralization, Democracy, Decarbonization and Digitalization;
- Relation to five main SPEN strategies or the 5S: Save, Shave, Share, Shine and Scale; and
- Relation to the goals in the SPEN framework: Energy and Environment, Economic, Indoor Environmental Quality (IEQ), Social and Smartness and Flexibility.

There are categorically inclusive aspects which means that the indicators can get multiple different tags (for example for Domain of sustainability, Life cycle stage, Scale, Authority, 5D, 5S, SPEN goals), while others are exclusive (for example Functionality, Type) meaning, that each indicator can have one tag. Functionality notation accounts for customization and prioritization, for when not all KPIs are relevant or measurable there needs to be a slight room for changes.

To calculate the KPI diversities, a well-known diversity index is used. The Shannon-diversity index is a broadly used diversity index in the field of ecology. It was first proposed by Claude Shannon [32], to quantify the entropy in strings of text. Here it is used to define the diversity of different KPI groups (grouped by indicator sets). Shannon diversity index is calculated as stated in Equation (1):

$$H' = \sum_{i=1}^{N} P_i \, \ln p_i \tag{1}$$

where $p_i$ is the proportion of indicators belonging to the i-th tag structure for the indicator set in question. DAGs are created to map the different indicators related to the drivers present for each performance to be measured in SPEN. Expert knowledge was used to define cause and effect relations between indicators of different drivers.

## 4. Results

As result of the application of the methodology described in Section 3, five KPI categories were identified and are defined as shown below:

- **Energy and Environmental,** which address overall energy and environmental performance, matching factors between load and on-site renewable generation and grid interaction;
- **Economic,** addressing capital costs and operational costs;
- **Indoor Environmental Quality (IEQ),** addressing thermal and visual comfort, as well as indoor air quality;
- **Social,** which addresses the aspects of equity, community and human outcomes; and
- **Smartness and Flexibility,** addressing the ability to be smartly managed.

Table 1 summarizes the 38 core indicators selected for the SPEN evaluation framework organised per category and sub-category. Details of each key indicators are described in the following sub-sections.

**Table 1.** Key Performance Indicators defined in the SPEN key performance categories.

| Category | Sub-Category | Key Performance Indicator |
|---|---|---|
| Energy and Environmental Performance | Overall Performance | Non-renewable primary energy balance<br>Renewable energy ratio |
| | Matching factor | Grid Purchase factor<br>Load cover factor/Self-generation<br>Supply cover factor/Self-consumption |
| | Grid interaction factors | Net energy/Net power<br>Peak delivered/exported power<br>Connection capacity credit |
| | Environmental balance | Total greenhouse gas emissions |
| Economic Performance | Capital costs | Investment costs<br>Share of investments covered by grants |
| | Operational costs | Maintenance-related costs<br>Requirement-related costs<br>Operation-related costs<br>Other costs |
| | Overall performance | Net Present Value<br>Internal Rate of Return<br>Economic Value Added<br>Payback Period<br>nZEB Cost Comparison |
| Indoor Environmental Quality | Indoor Air Quality | Carbon Dioxide ($CO_2$) |
| | Thermal comfort | Predicted Mean Vote (PMV)<br>Predicted Percentage Dissatisfied (PPD)<br>Temperature (T)<br>Relative Humidity (RH) |
| | Visual comfort | Illuminance<br>Daylight factor |
| | Acoustics comfort | Sound Pressure Level |

**Table 1.** *Cont.*

| Category | Sub-Category | Key Performance Indicator |
|---|---|---|
| Social Performance | Equity | Access to services<br>Affordability of energy<br>Affordability of housing<br>Democratic legitimacy<br>Living conditions |
| | Community | Social cohesion |
| | People | Personal safety<br>Energy consciousness |
| Smartness and Flexibility | Flexibility | Flexibility index |
| | Smartness | Smartness Readiness Indicator (SRI) |

The individual indicator sets have the following diversity indices in Table 2. The most evenly distributed diverse KPI variation can have a Shannon index of 2.30, which is the maximum value we can get from each indicator set. Considering the fact that every tag is represented at least once, and using a threshold value of 2.00 by Shannon indices, it can be stated that each indicator set is sufficiently diverse.

**Table 2.** Shannon index of each indicator set.

| Indicator Set | Shannon Index |
|---|---|
| Energy and Environment | 2.18 |
| Economic | 2.03 |
| Indoor Environmental Quality | 2.27 |
| Social | 2.23 |
| Smartness and Flexibility | 2.23 |

Causal DAG is created for all of the project goals. For the sake of simplification, these goals are presented by their DAG handles shown in Table 3. An overview of the DAG shows that for every main goal, there are at least two different impact chains, or in other words, there are at least two arrows pointing towards the goal from the outer circle in Figure 5. The five main goals are presented in the inside of the circle. The size of the nodes represents the number of arrows pointing to the node. The more inbound arrows are, the bigger the nodes. Edge colours are inherited from the target nodes at the end of each causal chain. Since for every goal there are multiple ways and multiple considerations considered, it is ensured that the risk of anchoring biases and the overreliance of available data is mitigated.

**Table 3.** Different SPEN goals categorised by the relevant key performance categories.

| Key Performance Category | 5D | 5S | SPEN Framework | DAG Handles SPEN Goals |
|---|---|---|---|---|
| Energy and Environment | Design<br>Decarbonisation<br>Decentralization | Save<br>Shave<br>Share | Self-consumption<br>GHG emissions | Decarbonisation |
| Economic | Design | Save<br>Scale | Cost efficiency<br>Self-consumption | Save |
| Indoor Environmental Quality | Democracy<br>Design | Shine | IEQ<br>Occupant satisfaction | Design |

**Table 3.** *Cont.*

| Key Performance Category | 5D | 5S | SPEN Framework | DAG Handles SPEN Goals |
|---|---|---|---|---|
| Social | Decentralization Democracy | Shine Share Save Scale | Social factors Occupant satisfaction | Democracy |
| Smartness and Flexibility | Digitalization Decentralization | Shave Share | Self-consumption GHG emissions | Digitalization & Decentralization |

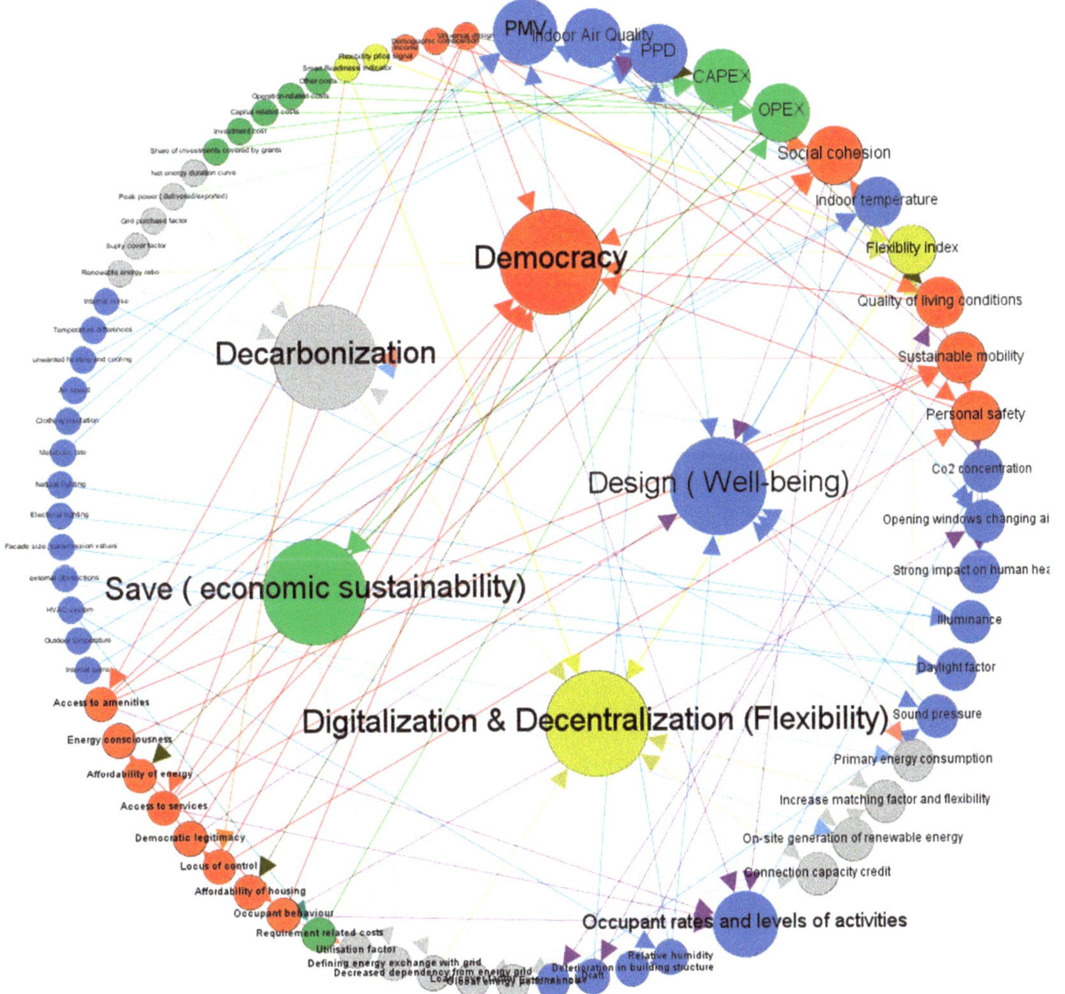


**Figure 5.** Complete causal DAG of the SPEN evaluation framework.

*4.1. Energy and Environmental Performance KPIs*

The set of indicators selected which characterizes the energy and environmental performance of each neighbourhood and their interaction with the connected energy networks follows the methodology of the Energy Performance of new and existing Buildings (EPB) described in the ISO-52000 standards [33] based on primary energy balance. In general terms, the overall energy performance of a building, by measurement or calculation, should be based on hourly or sub-hourly values of the different energy carrier flows in the buildings and by the exchanged energy (delivered and exported energy) with the energy networks in their broad concept (electricity, thermal energy with district heating and cooling networks, natural gas, biomass, etc.). Sub-categories for the energy and environmental assessment are depicted in Figure 6.

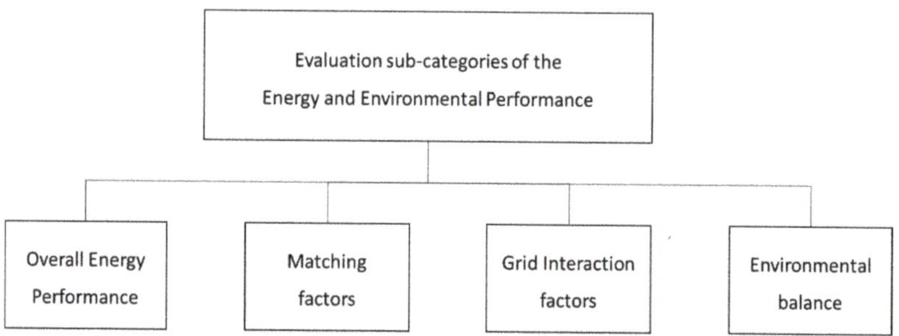


**Figure 6.** Evaluation sub-categories of the energy and environmental performance.

### 4.1.1. Overall Energy Performance

The overall energy performance of a system is calculated as the balance at the assessment boundary of the weighted delivered energy and weighted exported energy. The delivered energy is required to cover the energy demand of the considered neighbourhood, including the on-site generated energy, which can potentially be exported if not used within the neighbourhood. To describe the overall performance, two main indicators are selected. The main one is the non-renewable primary energy balance which weights the delivered and exported energy. If this balance is lower than zero, it means that it is a positive energy system. The other main indicator is the Renewable Energy Ratio which represents the share of renewable energy in the system.

- **Non-Renewable Primary Energy Balance:** This indicator takes into consideration all types of energy used and generated in the neighbourhood, and the exchange with the energy grids. It is calculated by summing all delivered and exported energy for all energy carries into a single indicator with the corresponding non-renewable primary energy conversion factors. In the framework of syn.ikia, weighting or conversion factors for exported energy should be selected based on the resources avoided from the external grid, which is equivalent to "Step B" stated in ISO-52000. This means that, for example, the values of the delivered and exported weighting factors for electricity are commonly considered to be equal.
- **Renewable Energy Ratio:** RER is the percentage share of energy from renewable sources in the total energy use. The share of renewable energy is defined by the Renewable Energy Ratio (RER), which is calculated relative to all energy use in the building, in terms of total primary energy and accounting for all the renewable energy sources. These include solar thermal, solar electricity, wind and hydroelectricity, renewable energy captured from ambient/ground heat sources by heat pumps and free cooling and renewable fuels [34].

### 4.1.2. Matching Factors

Load match factors describe the degree of the utilization of on-site energy generation related to the energy use in the neighbourhood. These factors characterize the direct use of energy generated inside the assessment boundary over a period and time (e.g., a day, a month or a year). Their calculation should be done on sub-hourly or hourly basis to characterize correctly the simultaneous use of on-site produced energy and the energy exchanged with the grid [2,35]. In the literature, the same concept has received different names. Two complementary indexes have been used: the load cover factor [2] or self-generation [36] and the supply cover factor [2,34] or self-consumption [36]. This can also be complemented with a third indicator:grid delivered factor or grid purchase ratio [37]. In case the energy use represents the useful energy demand, the grid purchase factor is a more reliable indicator and allows a fairer comparison of different systems, particularly if local electric and thermal storage are charged with renewables and/or the efficiencies of the compared systems differs.

- **Load Cover Factor/Self-Generation:** The load cover factor is the relation between the energy produced on-site and directly used and the total electric energy use. In ISO-52000, this factor is named use matching fraction.
- **Supply Cover Factor/Self-Consumption:** The supply cover factor is the relation between the energy produced on-site and directly used and the total on-site produced energy. In ISO-52000, this factor is named the production matching fraction.
- **Grid Delivered Factor:** The grid delivered factor is the relation between the energy delivered from the grid and the total energy used by the system over a time period. It characterizes the dependency of the neighbourhood of the grid [37].

### 4.1.3. Grid Interaction Factors

Grid interaction indicators are based on the net energy which represents the electricity interaction between the neighbourhood and the grid, per energy carrier. For a proper analysis of grid interaction, sub hourly resolution data is required (recommended in the range of 1–5 min and 15 min as a maximum) as there is a relatively high impact due to time averaging effects [38].

- **Net Energy/Net Power:** Net energy allows one to assess the interaction of a system with the energy grids over a certain period: a day, a month or a year. In doing that, it is useful to represent the net energy using a duration curve, colored carpet plots and/or box plots [2]. This kind of visual representations allows for an immediate comprehension of the distribution of power and the differences between alternative solutions. Figure 7 shows schematically the net energy duration curve. It should be noted that the red area of the net load duration curve represents the net delivered energy. In the case of a yearly duration curve, the red area of the duration curve is equal to annual delivered energy, while the green area is equal to annual exported energy. In coherence with the definition of SPEN and the ISO52000 set of standards, we refer to net energy exchange as a result of an energy balance considering on-site/nearby generated energy to cover the EPD energy use. If parts of the energy uses of the building and neighbourhood are discarded in the energy assessment, actual metered grid interaction will differ from the calculated one, as represented schematically in Figure 7.
- **Peak Delivered/Peak Exported Power:** The peak delivered and peak exported power KPIs are the extreme values of the net duration curve. The maximum positive value is the peak delivered, while the maximum negative value is the peak exported.
- **Connection Capacity Credit:** The connection capacity credit, or power reduction potential [39], is defined as the percentage of grid connection capacity that could be saved compared to a reference case [40,41].

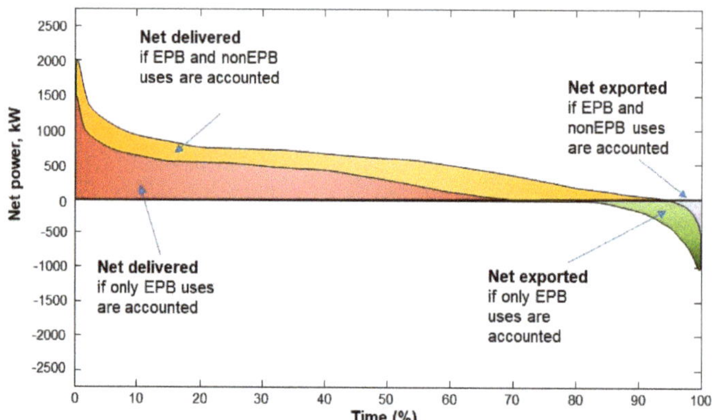



**Figure 7.** Net energy duration curve considering EPB energy use and non-EPB energy use in a neighbourhood: conceptual scheme. Source: IREC.

### 4.1.4. Overall Environmental Balance

Overall Environmental balance is assessed using the total greenhouse gas emissions as the indicator.

- **Total Greenhouse Gas Emissions:** It is calculated in a similar way that the primary energy balance and takes into consideration all types of energy used and generated by the system, and the exchange with the energy networks. It is calculated summing up all delivered and exported energy for all energy carries into a single indicator with the emissions of the delivered and exported energy carriers as weighting factors.

### 4.2. Economic Performance

The set of indicators for demonstrating economic performance is selected from the perspective of building owners and investors. SPENs are more expensive than traditional projects and the main barrier for SPEN development is the access to adequate funding and business models [42]. Yet, PEDs and SPENs also hold the potential for fostering economic sustainability due to cost efficiency and self-consumption. To the potential investor who is considering whether to invest in a SPEN, or to the building owner who wishes to track and reflect the savings from the building level to a neighbourhood scale, the following categories of indicators are recommended to be accounted for: capital costs, operational costs and overall performance, as reflected in Figure 8.

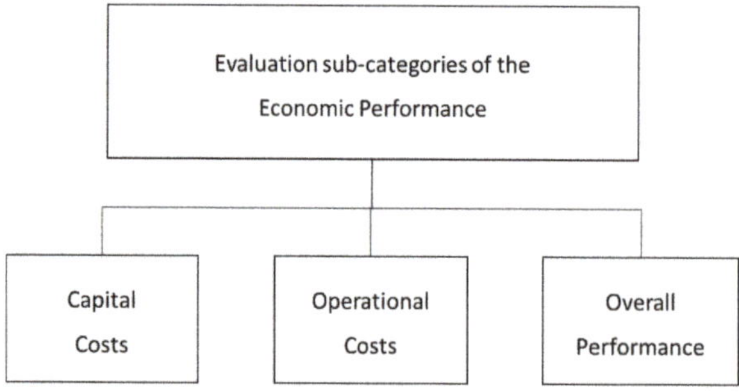



**Figure 8.** Evaluation sub-categories of the economic performance.

### 4.2.1. Capital Costs

Capital costs in the SPEN evaluation framework refer to complete building construction cost and the cost of assets or items that are purchased or implemented with the aim of improving the energy efficient aspects of the system [41]. As stated earlier, such assets or items can include but are not limited to multi-functional façade elements with integrating photovoltaic and solar thermal, heat pumps, thermal storages and batteries.

- **Investment Costs** ($€/m^2$): This indicator calculates the costs of the newly built or refurbished building, assets or items and is defined as cumulated payments until the initial operation of the system.
- **Share of Investments Covered by Grants** ($€/m^2$): This indicator considers any grants or subsidies that should be accounted as capital costs in order to reflect a truly market-based approach in evaluating the cost efficiency of SPENs.

### 4.2.2. Operational Costs

In the operation and maintenance of SPENS, respective operational costs are incurred, from maintenance to repairs and replacements over a period of time. These costs can vary for each year [41].

- **Maintenance-Related Costs** ($€/m^2/year$): This indicator encompasses depreciation, interests, repairs and replacements of those assets or items purchased or implemented to improve the energy efficiency aspects of the system.
- **Requirement-Related Costs** ($€/m^2/year$): This indicator is defined as power and fuel costs, costs for operating resources and in some cases external costs.
- **Operation-Related Costs** ($€/m^2/year$): This indicator relates to the costs of using the installation.
- **Other Costs** ($€/m^2/year$): This indicator captures other costs such as the cost of insurance.

### 4.2.3. Overall Economic Performance

This sub-category of overall economic performance represents important factors in the decision-making in real estate markets (new built and renovation) from the perspective of building owners and investors. They provide an evaluation of the relative benefits of a particular choice of investment. They summarize both the capital costs and the operational costs together with possible sources of income in a single indicator. Within this sub-category of KPIs, the Net Present Value is the one that is considered the most reliable [43]. In some cases, especially when liquidity is a limiting factor, other KPIs, such as the Payback Period, might be more relevant to the building owner or investor.

- **Net Present Value** (€): The Net Present Value (NPV) is computed as the difference between the investment and the discounted cash flows related to an investment. In the context of SPENs, the cash flows can be represented by the yearly savings obtained by entering the project. These savings can be discounted using a risk-adjusted rate of return to provide an estimate of the value of these savings as if the investors would obtain them at the same moment when the investment occurs. The discount rate needs to be defined using available ones employed in similar projects or recovered from the stock market.
- **Internal Rate of Return:** The Internal Rate of Return (IRR) is defined as the discount rate that makes the current value of savings equal to the initial investment.
- **Economic Value Added** (€):The Economic Value Added is a quick evaluation measure that can be computed as the difference between the yearly savings and the minimum required savings.
- **Payback Period** (year): The Payback Period is the number of years it takes before the cumulative savings equals the initial investment.
- **nZEB Cost Comparison** (%): The nZEB Cost Comparison is computed as the ratio between the total cost of the respective investment and its nZEB alternative. The

calculation period should cover the expected lifetime of the SPEN and the reference, e.g., 50 years.

### 4.3. Indoor Environmental Quality

People spend approximately 90% of their time in indoor environments [44]. Over the last decades, an abundant number of studies have shown that the indoor environmental quality (IEQ) has a significant impact on human health and wellbeing [45]. IEQ refers to the quality of a building's environment with respect to wellbeing and health of the building occupants and is determined by many factors such as indoor thermal environment, air quality and lighting and acoustics [46]. Well designed and implemented plus energy buildings and neighbourhoods can bring multiple benefits, including improvements in air quality, health comfort and productivity. It is therefore essential to ensure that the IEQ positively contributes to realising these benefits.

This section aims at developing an approach to assess the IEQ of plus energy buildings by focusing on the main factors that determine the indoor environment (see Figure 9). A common approach to assess IEQ can help highlighting potential areas for improvement and provide useful feedback to building professionals and value chain actors, including designers, developers, facilities managers and property agents. The evaluation framework has been designed with the objectives of user friendliness, quality, reliability and economic feasibility. It is built on existing methodologies, frameworks, indexes and certification schemes such as Level(s) [47], CBE Survey [48], TAIL [49], DEQI [50], WELL [51], IEQ-Compass [52]. It also complies with the EN Standard 16798 [53,54]. The evaluation framework can be used at several stages of the life cycle of the buildings in a SPEN. The predicted IEQ characteristics of the buildings are explored at the design phase through calculations and simulations, while the actual IEQ is assessed during the operational phase through on-site measurements, checklists, and questionnaire surveys. This approach allows to determine whether the SPEN meet their design objectives but also make a link between design and operational performance.

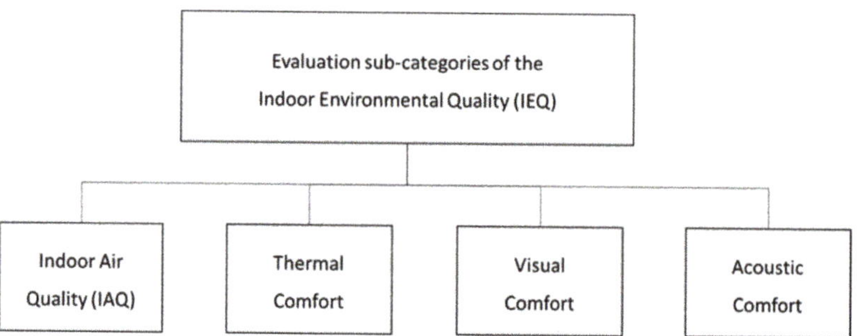


**Figure 9.** Evaluation sub-categories of the IEQ category.

### 4.3.1. Indoor Air Quality (IAQ)

Adequate indoor air quality is the air without harmful concentrations of contaminants [55]. Studies have linked poor indoor air quality with adverse health effects such as asthma, eczema and allergic diseases [56]. Carbon dioxide is of the most well-known contaminants and is a good proxy of the indoor air quality as it can provide an indication of the ventilation rate in a space. Therefore, $CO_2$ (in units of ppm) has been selected as the KPI of the IAQ and its concentration ranges will be used to evaluate the indoor air quality according to the four categories specified in EN ISO 16798-1-2019.

### 4.3.2. Thermal Comfort

According to the EN ISO 7730, "thermal comfort is that condition of mind which expresses satisfaction with the thermal environment". The level of occupant's thermal comfort is often expressed in percentage of the number of people who are satisfied or dissatisfied with the thermal conditions. The most commonly used indexes are the predicted mean vote (PMV) and the predicted percentage dissatisfied (PPD)—these will be KPIs of the thermal environment. The calculation formulas for the PMV and PPD indexes are in line with the ISO 7730 and ASHRAE Standard 55. Air temperature (°C) and relative humidity (%) will be additional KPIs of the thermal environment of buildings without mechanical cooling. To evaluate the thermal environment, the percentage of time that temperatures are out of the ranges specified in the categories of EN 16798, should be estimated for buildings with and without cooling systems for the heating and cooling seasons.

### 4.3.3. Lighting and Visual Comfort

According to the EN12665, visual comfort is defined as "a subjective condition of visual well-being induced by the visual environment". A good visual environment (e.g., adequate levels of natural and artificial lighting, reduced glare, etc.) can add to the well-being and productivity of the building occupants [57]. Illuminance is the total amount of light delivered on a surface by either natural daylight or electrical fitting. In this project, the illuminance (lux) and the daylight factor (%) will be measured and simulated to evaluate the visual environment and will serve as the KPIs of the lighting and visual comfort. Daylight factor is a metric expressing, as a percentage, the amount of daylight that is available in a room in comparison to the amount of daylight available outside under overcast sky conditions [57]. The daylight factor depends on the size, the transmission properties of the façade, the size and shape of the space as well as the extent to which external structures obscure the view of the sky.

### 4.3.4. Acoustic Comfort

Acoustic comfort includes the protection of building occupants from noise in order to provide a suitable acoustic environment for the designed human activity [58]. Depending on the levels of noise, it can cause annoyance, hearing damage or interference to speech intelligibility [59]. The acoustic environment should be designed to avoid these harmful effects and the criteria used to ask for an acceptable c environment are expressed in sound levels decibels (dB), noise rating (NR) or noise criteria (NC). To determine the quality levels of acoustic comfort in the living room, the percentage of hours that the level of acoustics exceed noise levels defined in the categories specified in EN 16798 will be estimated. The sound pressure level (dB(A)) will serve as the KPI of the acoustic comfort.

### 4.4. Social Performance

There are no standards on how to monitor social sustainability, which is due to inconsistencies in its definition [60]. Defining what social includes and what it does not is bound to political and contextual factors [61]. Social performance in the SPEN context is defined as the fidelity of development with human and societal values. This evolution should foster an environment that achieves reconciling cohabitation and heterogeneity, fostering cohesive community practices and improving in quality of life for. To achieve this, social performance is assessed on three pillars simultaneously (see Figure 10).

- **Equity:** assessment of the fair, just and legitimate functioning of the community.
- **Community:** assessment of the ability of the community to maintain itself and thrive.
- **People:** assessment of human experiences, behaviour and outcomes.

Some methodological adjustments are necessary when evaluating social performance [62]. First, social KPIs vary most across different scales: on national levels demographic, systemic variables are prevalent, while on hyperlocal levels focus more on social interactions and quality of life. The neighbourhood scale is a mix of both, since it is place-based, but also requires some form of institutionalization in the PED context [63]. Second, it is not always apparent, whether a specific result for an indicator is good or bad, plus, in many cases this is location-dependent, meaning universal benchmarks are rarely feasible [60,64]. Third, social performance should be measured both as an objective variable and as the way people feel about it to account for varying user experiences [65]. Certain KPIs must couple objective and subjective components to avoid institutionalizing injustices of unconsidered human experiences [66]. The data collection for subjective experiences, however, must be carefully designed to avoid a collection of subjective interpretations—which can result in inconsistent responses. Finally, the distribution of social performance must also be monitored to avoid obscuring disparities and discrimination among different social groups [61].

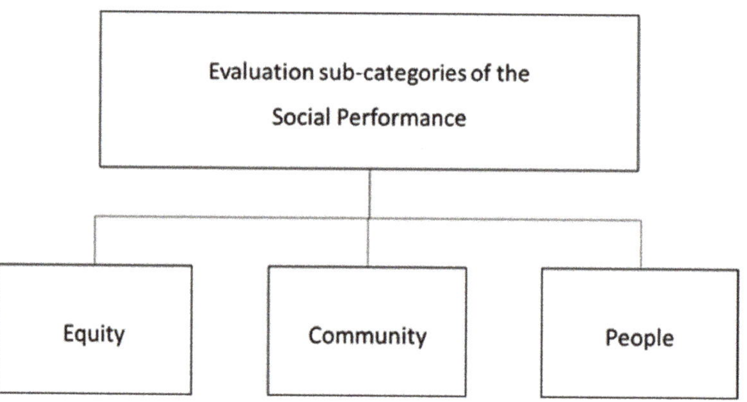


**Figure 10.** Evaluation sub-categories of the social performance category.

### 4.4.1. Equity

Equity indicators describe the fair, just, legitimate functioning of the SPEN. It refers to various aspects of justice, including spatial and procedural, referring to the distribution of services (such as education, green spaces, adequate and affordable housing, public transportation), and participation on important issues, especially where valuable infrastructure (here, energy assets) are shared. It includes just, equitable, accountable distribution of influence, blind to all aspects of identity or personal resources. Core KPIs are:

- **Access to Services** measures whether services of general interest (education, groceries, healthcare, green, etc. as discussed in EC Com (2011) 900) are in walkable distance to all households. It is a GIS-based assessment through the aggregation of distances evaluated against service-specific thresholds for each service and each household. This metric indicates both walkability—for an environmental perspective [67]—and spatial justice—for a social perspective [68].
- **Affordability of Energy** is the adoption the standardised European "arrears" and "share of income spent on energy" indicators for energy poverty developed by the European Energy Poverty Observatory (EPOV) [69]. It is measured through a household survey and reflects the wider goal of providing just sustainable transition for all [70].
- **Affordability of Housing** is measuring, in a household survey the share of people with a housing cost overburden, and mapping the income required to afford housing in the SPEN. This is partially in line with the European standard of relating housing costs to income [71–73], but considers more items on the cost side to fully reflect real cost of housing—notably mortgage principals. Also, the second metric of the KPI indicates whether the SPEN is gated or affordable for the larger population.
- **Democratic Legitimacy** is measured by two sets of criteria: objective and subjective. Objective criteria are measured on the process and content of stakeholder consultation [74], while the subjective part is a survey of participants on their experiences of the process [72]. Legitimacy is critical to ensure the principle of subsidiarity [75], and a fair consideration of individual interests in collective and top-down decisions [36].

- **Living Conditions** are the adoption of overcrowding and common poor living condition "red flags" monitored by Eurostat, which in turn refers to SDGs 1, 6 and 11 [67,73]. These are measured in household surveys as they are determinants of social, health-related and environmental outcomes [67,69,71,73].

In addition to the core KPIs presented here, complementary KPIs that may be included in this category are access to amenities [68], sustainable mobility and accessibility to universal design [76].

### 4.4.2. Community

Community indicators describe the resilience, and self-reinforcing quality of local communities. On the one hand, this refers to maintenance of social networks, including absorbing newcomers and engaging with existing members. On the other hand, this refers to social capital exchange in social networks, including the use of public spaces and other channels for meaningful interactions, conflict resolution, and supporting one another.

- **Social Cohesion** addresses the existence and the conditions for strong social networks, formed on trust-based bonds, with a capacity to absorb and build on diversity [72,77,78]. The indicator has a subjective, normative component that evaluates personal resilience attributed to belonging in a household survey [72,78]. It also has descriptive component, in a form of a checklist of environmental features that can support social cohesion [71,79].

### 4.4.3. People

People indicators refer to social performance measured on individuals and describing personal, human conditions. These cover human needs, like health, employment, education, security, and quality of life metrics, such as wellbeing, happiness and comfort. Additionally, people indicators include environmental determinants for both of the above. Lastly, they describe how sustainably people inhabit their SPEN, how they behave and interact with their environment, and its resources.

- **Personal Safety** refers to the goal of providing safe, non-intrusive public and shared spaces eliminating deterrents of walking and staying outdoors, especially for women [68]. The indicator has a subjective, normative component that is the adoption of standard Eurostat metrics in a household survey [73]. It also has a descriptive component, in a form of a checklist of environmental features that are associated with perceived and real safety in public space.
- **Energy Consciousness** describes the behavioural determinants of energy use, which is crucial to eliminate occupant-centric barriers to coordinated, environmentally conscious energy management [63]. The indicator is measured through a household survey. The survey extracts personal drivers behind environmental, energy, and technology-related decisions based on common behavioral models [80,81]. These helps classify the main drivers per social group and guidelines are provided how to respond to specific driver classes.

### 4.5. Smartness and Flexibility

Smartness and flexibility refers to the ability of the built environment to manage its energy demand and local generation according to the climate conditions, user needs and preferences and grid requirements. For its assessment, two sub-categories are defined: flexibility and smartness (see Figure 11). The Flexibility Index and the Smart Readiness Indicator (SRI) are proposed as KPIs, respectively.

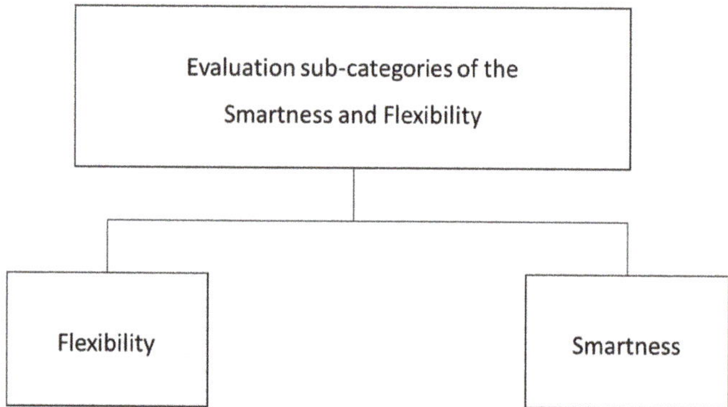



**Figure 11.** Evaluation sub-categories of the smartness and flexibility.

### 4.5.1. Flexibility

The flexibility index is proposed to evaluate flexibility in SPENs. The flexibility index has been developed during the past years and several publications explain it in details [17,75]. The flexibility index is related to the ability of a system to react at a variable price signal and, thus, optimize the energy usage of a given system to minimize the energy cost.

- **The flexibility index** is defined as the monetary savings that can be achieved by adopting a flexible energy usage, for a given price-signal [82]. For example, given a price-signal, a building or a neighbourhood obtaining a Flexibility Index of 0.1 means that the building or neighbourhood is able to save 10% of its energy costs, by applying energy flexibility, for the given price signal. It shall be noted that, if the price reflects the amount of $CO_2$ emissions of the energy mix used as energy source, the flexibility index is also representing the savings of $CO_2$ emissions obtained through the smart controller.

### 4.5.2. Smartness

The SRI [83] at the building level has been adopted by the EU as the main measure to evaluate how smart-ready buildings are (https://smartreadinessindicator.eu/ (accessed on 16 July 2021)). According to the executive summary [84] on the "Smart Readiness Indicators", the aim of the SRI is to "make the added value of building smartness more tangible for building users, owners, tenants and smart service providers". The roll-out scheme for the SRI implementation across the EU procedure was published in October 2020 [85] The definition of Smartness in buildings can be adapted to the neighbourhood level.

- **Smartness Readiness Indicator of a SPEN** refers to the ability of a neighbourhood (namely, its systems and buildings) to sense, interpret, communicate and actively respond in an efficient manner to changing conditions in relation to the operation of technical building systems or the external environment and to demands from the occupants and the users of the different buildings and services. The readiness of a neighbourhood to be smart is related to three aspects: adapting in response to the needs of the occupants and users, facilitating the maintenance and efficient operation processes and adapting in response to (price) signals from the grid. From the practical point of view, it is suggested to compute the SRI for each building of a SPEN, singularly.

## 5. Discussion

This paper presents two main contributions to the existing literature. The first one is to put forth a consistent definition of Sustainable Plus Energy Neighbourhood (SPEN) with the objective to make an active contribution to the ongoing debate on a common understanding of what a Positive Energy District (PED) is and how it should be evaluated. Acknowledging the positive benefits to act at larger scale than at individual buildings, a SPEN is defined as a group of interconnected buildings with associated infrastructure, located within both a confined functional area and/or virtual boundary. A SPEN aims to reduce its direct and indirect energy use towards zero over an adopted complete year with an increased use and production of renewable energy. Beyond the consideration of the energy balance, several other aspects need to be considered to achieve a successfully sustainable PED. Then, additionally, the definition of a SPEN covers the following five main objectives:

- The primary energy, net-zero greenhouse gas emissions and carbon footprint reduction;
- The active management of annual site or regional surplus production of renewable energy and power performance (self-consumption, peak shaving, etc.) through smart management and energy flexibility;
- The cost efficiency and economic sustainability according to a life cycle assessment;
- An improved indoor environment for well-being of the inhabitants; and
- The social inclusiveness, interaction and empowerment related to co-use, shared services and affordable living.

The second major contribution is to present an evaluation framework for the assessment of SPENs, to be applied both during the design phase and the operational phase to monitor their performance. The evaluation framework defines five main categories and the KPIs which are essential for the evaluation of SPENs are described with the rationale of their selection. The selection of the main assessed categories and KPIs have been based on a holistic and exhaustive methodology which highlights the multiple dimensions when addressing sustainability in districts as moving beyond the traditional and strict building energy assessment. The indicators set for each category were selected with the objective to be diverse enough and to represent the SPEN goals in a balanced and integrated way. Methodology is based on a tagging structure, DAGs and expert knowledge. A total number of 38 KPIs that allow addressing the multidimensionality nature of SPENs are described and distributed in five categories which are:

- Energy and Environmental: addressing overall energy and environmental performance, matching factors between load and on-site renewable generation and grid interaction;
- Economic: addressing capital costs and operational costs;
- Indoor Environmental Quality (IEQ): addressing thermal and visual comfort, as well as indoor air quality;
- Social indicators: addressing the aspects of equity, community and people; and
- Smartness and Energy Flexibility: addressing the ability of manage energy in clusters of buildings according sustainable objectives.

Our study provides additional support and considerable insight to large scale deployment of SPENs and PEDs. The next steps and further work will include the testing and validation of the proposed assessment framework and the indicators in four real projects across Europe, both in the integrated design phase and the operational phase. The projects are real estate developments pursuing the SPEN's objectives in four different climate zones (Subartic, Marine, Mediterranean and Continental) with different housing contexts [1]. Our research can be a useful aid for the design, construction and post-occupancy evaluation of PEDs in an holistic way.

**Author Contributions:** Conceptualization, J.S., I.A. and N.G.; methodology, J.S., A.R., Á.M., P.V.D., C.C., I.A. and D.C.; investigation, M.T., P.C., D.C., V.B., R.B., P.V.D., C.M., C.C. and Z.T.; writing—original draft preparation, M.T., P.C., V.B., P.V.D., D.C. and C.C.; writing—review and editing, J.S., I.A., J.P., A.R. and N.G.; visualization, P.C.; supervision, J.S.; funding acquisition, N.G. All authors have read and agreed to the published version of the manuscript.

**Funding:** The research conducted in this paper has received funding from the European Union's Horizon 2020 research and innovation programme under grant agreement No 869918, SYN.IKIA

Project. Paolo Civiero acknowledges the funding received from the European Union's Horizon 2020 research and innovation programme under the Marie Skłodowska-Curie grant agreement No 712949 (TECNIOspring PLUS) and the Agency for Business Competitiveness of the Government of Catalonia.

**Institutional Review Board Statement:** Not applicable.

**Informed Consent Statement:** Not applicable.

**Data Availability Statement:** Not applicable.

**Conflicts of Interest:** The authors declare no conflict of interest.

## Abbreviations

| | |
|---|---|
| CEN | European committee for standarisation |
| CHP | Combined heat and power |
| DAG | Directed acyclic graph |
| DER | Distributed energy resources |
| DG | Distributed generation |
| DR | Demand response |
| EED | Energy efficiency directive |
| EPB | Energy performance of a building |
| EPBD | Energy performance of buildings directive |
| EV | Electrical vehicle |
| GHG | Greenhouse gas |
| IAQ | Indoor air quality |
| ICT | Information and communications technology |
| IEQ | Indoor environmental quality |
| IRR | Internal rate of return |
| KPI | Key performance indicator |
| M&E | Measurement and evaluation |
| NC | Noise criteria |
| NPV | Net present value |
| NR | Noise rating |
| nZEB | Nearly zero energy building |
| PEB | Positive energy building |
| PED | Positive energy district |
| PMV | Predicted mean vote |
| PPD | Predicted percentage dissatisfied |
| RER | Renewable energy ratio |
| RES | Renewable energy resources |
| RH | Relative humidity |
| SDG | Sustainable development goal |
| SPEN | Sustainable positive energy neighbourhood |
| SRI | Smart readiness indicator |
| WEC | World energy council |

## References

1. NTNU. Syn.Ikia—Sustainable Plus Energy Neighbourhoods-EU H2020 G.A. No. 841850. Available online: https://www.synikia.eu/ (accessed on 10 June 2021).
2. Salom, J.; Marszal, A.J.; Widén, J.; Candanedo, J.; Lindberg, K.B. Analysis of load match and grid interaction indicators in net zero energy buildings with simulated and monitored data. *Appl. Energy* **2014**, *136*, 119–131. [CrossRef]
3. Marszal-Pomianowska, A.; Johra, H.; Knotzer, A.; Salom, J.; Péan, T.; Jensen, S.Ø.; Mlecnik, E.; Kazmi, H.; Pernet, R.; Klein, K.; et al. *International Energy Agency—Principles of Energy Flexible Buildings: Energy in Buildings and Communities Programme Annex 67 Energy Flexible Buildings*; Danish Technological Institute: Taastrup, Denmark, 2019.
4. Wiik, M.K.; Fufa, S.M.; Baer, D.; Sartori, I.; Andresen, I. *The Zen Definition—A Guideline for the Zen Pilot Areas*; Research Report; SINTEF Akademisk Forlag: Trondheim, Norway, 2018; No. 11.
5. Proposal for the European Partnership Driving Urban Transitions Draft Proposal for a European Partnership under Horizon Europe Driving Urban Transitions to a Sustainable Future (DUT). Available online: https://ec.europa.eu/info/files/european-partnership-driving-urban-transitions-sustainable-future-dut_en (accessed on 16 July 2021).

6. EU SETIS SET-Plan Temporary Working Group 3.2. SET-Plan ACTION n°3.2 Implementation Plan. 2018. Available online: http://docplayer.net/99748225-Set-plan-action-n-3-2-implementation-plan.html (accessed on 17 July 2021).

7. EERA Joint Programme Smart Cities. Available online: https://www.eera-sc.eu/ (accessed on 10 June 2021).

8. European Innovation Partnership on Smart Cities and Communities. Available online: https://smart-cities-marketplace.ec.europa.eu/ (accessed on 10 June 2021).

9. European Regions Research and Innovation Network. Available online: https://errin.eu/ (accessed on 10 June 2021).

10. Eurocities. Available online: http://www.eurocities.eu/ (accessed on 10 June 2021).

11. International Energy Agency EBC Annex 83 PEDs. Available online: https://annex83.iea-ebc.org/ (accessed on 10 June 2021).

12. European Cooperation in Science and Technology Action. Available online: https://www.cost.eu/ (accessed on 10 June 2021).

13. Polly, B.; Pless, S.; Houssainy, S.; Torcellini, P.; Livingood, W.; Zaleski, S.; Jungclaus, M.; Hootman, T.; Craig, M. *A Guide to Energy Master Planning of High-Performance Districts and Communities*; Technical Report; National Renewable Energy Lab. (NREL): Golden, CO, USA, 2020.

14. JPI Urban Europe/SET Plan Action 3.2. Europe Towards Positive Energy Districts. PED Booklet. 2020. Available online: https://jpi-urbaneurope.eu/ped/ (accessed on 10 June 2021).

15. The Smart Cities Information System (SCIS). Actions & Recommendations: Creating A Joint Vision for Peds (Positive Energy Districts). 2020. Available online: https://cityxchange.eu/eusew-2020-webinar-creating-a-joint-vision-for-positive-energy-districts/ (accessed on 16 July 2021).

16. van Dijk, D.; Hogeling, J. The new EN ISO 52000 family of standards to assess the energy performance of buildings put in practice. *E3S Web Conf.* **2019**, *111*. [CrossRef]

17. Stankuniene, G.; Streimikiene, D.; Kyriakopoulos, G.L. Systematic literature review on behavioral barriers of climate change mitigation in households. *Sustainability* **2020**, *12*, 7369. [CrossRef]

18. Yu, X.; Ma, S.; Cheng, K.; Kyriakopoulos, G.L. An evaluation system for sustainable urban space development based in green urbanism principles—A case study based on the Qin-Ba mountain area in China. *Sustainability* **2020**, *12*, 5703. [CrossRef]

19. Jensen, S.Ø.; Madsen, H.; Lopes, R.; Junker, R.G.; Aelenei, D.; Li, R.; Metzger, S.; Lindberg, K.B.; Marszal, A.J.; Kummert, M.; et al. *Annex 67: Energy Flexible Buildings—Energy Flexibility as a Key Asset in a Smart Building Future—Contribution of Annex 67 to the European Smart Building Initiatives*; EBC: Vienna, Austria, 2017; pp. 1–16. Available online: http://www.annex67.org/media/1470/position-paper-energy-flexibility-as-a-key-asset-i-a-smart-building-future.pdf (accessed on 16 July 2021).

20. Wang, N.; Phelan, P.E.; Gonzalez, J.; Harris, C.; Henze, G.P.; Hutchinson, R.; Langevin, J.; Lazarus, M.A.; Nelson, B.; Pyke, C.; et al. Ten questions concerning future buildings beyond zero energy and carbon neutrality. *Build. Environ.* **2017**, *119*, 169–182. [CrossRef]

21. Brozovsky, J.; Gustavsen, A.; Gaitani, N. Zero emission neighbourhoods and positive energy districts—A state-of-the-art review. *Sustain. Cities Soc.* **2021**, *72*, 103013. [CrossRef]

22. Salom, J.; Tamm, M. Methodology Framework for Plus Energy Buildings and Neighbourhoods. 2020. Available online: https://www.synikia.eu/wp-content/uploads/2020/12/D3.1_Methodology-framework-for-Plus-Energy-Buildings-and-Neighbourhoods.pdf (accessed on 16 July 2021).

23. Reith, A. Restore—WG5: Scale jumping. In Proceedings of the Presented at the Kick off Meeting, Larnaca, Cyprus, 2 March 2020.

24. Institut Català d'Energia, Smart Energy Communities: Insights into Its Structure and Latent Business Models. 2019. Available online: http://icaen.gencat.cat/web/.content/10_ICAEN/17_publicacions_informes/11_altres_publicacions/arxius/SmartEnergyCommunities.pdf (accessed on 16 July 2021).

25. Valdés, J. Arbitrariness in multidimensional energy security indicators. *Ecol. Econ.* **2018**, *145*, 263–273. [CrossRef]

26. Tversky, A.A.; Kahneman, D.D. Judgement under uncertainty: Heuristics and biases. *J. Exp. Soc. Psychol.* **1974**, *6*, 401–419. [CrossRef]

27. Adame, B.J. Training in the mitigation of anchoring bias: A test of the consider-the-opposite strategy. *Learn. Motiv.* **2016**, *53*, 36–48. [CrossRef]

28. Weaver, W. Science and complexity. *Am. Sci.* **1948**, *36*, 536–544.

29. Zeng, G.; Zeng, E. On the relationship between multicollinearity and separation in logistic regression. *Commun. Stat. Simul. Comput.* **2019**, 1–9. [CrossRef]

30. Pearl, J.; Dechter, R. Identifying independencies in causal models with feedback. *Acta Math. Appl. Sin.* **2000**, *23*, 299–310.

31. Tanguay, G.A.; Rajaonson, J.; Lefebvre, J.F.; Lanoie, P. Measuring the sustainability of cities: An analysis of the use of local indicators. *Ecol. Indic.* **2010**, *10*, 407–418. [CrossRef]

32. Shannon, C.E. A mathematical theory of communication. *Bell Syst. Tech. J.* **1948**, *27*, 379–423. [CrossRef]

33. *International Standard Overarching EPB Assessment*; ISO: Geneva, Switzerland, 2017.

34. Mazzarella, L. *Rehva Report 04: nZEB Technical Definition and System Boundaries for Nearly Zero Energy Buildings*; REHVA: Ixelles, Belgium, 2013; No. 4.

35. Cao, S.; Hasan, A.; Sirén, K. On-site energy matching indices for buildings with energy conversion, storage and hybrid grid connections. *Energy Build.* **2013**, *64*, 423–438. [CrossRef]

36. Brattebø, H.; Gustavsen, A.; Wiik, M.K.; Fufa, S.M.; Krogstie, J.; Ahlers, D.; Wyckmans, A.; Driscoll, P. *Zero Emission Neighbourhoodsin Smart Cities: Definition, Key Performance Indicators and Assessment Criteria*; Techincal Report; SINTEF Akademisk Forlag: Trondheim, Norway, 2018; Version 1.0, No. 7.

37. Haberl, R.; Haller, M.; Bamberger, E.; Reber, A. Hardware-In-the-loop tests on complete systems with heat pumps and PV for the supply of heat and electricity. In Proceedings of the 12th International Conference on Solar Energy for Buildings and Industry, Rapperswil, Switzerland, 10–13 September 2018; pp. 1–10. [CrossRef]

38. Salom, J.; Widén, J.; Candanedo, J.; Lindberg, K.B. Analysis of grid interaction indicators in net zero-energy buildings with sub-hourly collected data. *Adv. Build. Energy Res.* **2015**, *9*, 89–106. [CrossRef]

39. Verbruggen, B.; de Coninck, R.; Baetens, R.; Saelens, D.; Helsen, L.; Driesen, J. Grid impact indicators for active building simulation. In Proceedings of the ISGT 2011 (the Innovative Smart Grid Technologies), Anaheim, CA, USA, 17–19 January 2011. [CrossRef]

40. Tejero, A.; Ortiz, J.; Salom, J. Evaluation of occupancy impact in a residential multifamily nZEB through a high resolution stochastic model. In Proceedings of the 4th Building Simulation and Optimization Conference, Cambridge, UK, 11–12 September 2018.

41. Marijuán, G.; Etminan, G.; Möller, S. Smart cities information system—Key performance indicator guide version 2.0. In Proceedings of the 3rd IEEE Conference on Smart Cities and Innovative Systems, Marrakesh, Morocco, 21–24 October 2018.

42. Bossi, S.; Gollner, C.; Theierling, S. Towards 100 positive energy districts in Europe: Preliminary data analysis of 61 European cases. *Energies* **2020**, *13*, 6083. [CrossRef]

43. Brealey, R.A.; Myers, S.C.; Allen, F. *Principles of Corporate Finance*, 10th ed.; McGraw-Hill/Irwin: New York, NY, USA, 2011.

44. Klepeis, N.E.; Nelson, W.C.; Ott, W.R.; Robinson, J.P.; Switzer, P. The national human activity pattern survey (NHAPS): A resource for assessing exposure to environmental pollutants. *J. Expo. Sci. Environ. Epidemiol.* **2001**, *11*, 231–252. [CrossRef] [PubMed]

45. Al horr, Y.; Arif, M.; Katafygiotou, M.; Mazroei, A.; Kaushik, A.; Elsarrag, E. Impact of indoor environmental quality on occupant well-being and comfort: A review of the literature. *Int. J. Sustain. Built Environ.* **2016**, *5*, 1–11. [CrossRef]

46. NIOSH. Indoor Environmental Quality. Centers for DIsease Control and Prevention—The National Institute for Occupational Safety and Health (NIOSH). Available online: https://www.cdc.gov/niosh/topics/indoorenv/default.html (accessed on 16 July 2021).

47. Dodd, N.; Cordella, M.; Travreso, M.; Donatello, S. *Level(s)—A Common EU Framework of Core Sustainability Indicators for Office and Residential Buildings Part 3: How to Make Performance Assessments Using Level(s) (Beta v1.0)*; Publications Office of the EU: Luxembourg, 2017. [CrossRef]

48. Center for the Built Environment (CBE). Occupant Indoor Environmental Quality Survey. Available online: https://cbe.berkeley.edu/research/occupant-survey-and-building-benchmarking/ (accessed on 16 July 2021).

49. Wargocki, P.; Mandin, C.; Wei, W. ALDREN Methodology Note on Addressing Health and Wellbeing. 2020. Available online: https://aldren.eu/wp-content/uploads/2020/12/D2_4.pdf (accessed on 16 July 2021).

50. Laskari, M.; Karatasou, S.; Santamouris, M. A methodology for the determination of indoor environmental quality in residential buildings through the monitoring of fundamental environmental parameters: A proposed Dwelling Environmental Quality Index. *Indoor Built Environ.* **2017**, *26*, 813–827. [CrossRef]

51. WELL, The WELL Building Standard v1. 2019. Available online: https://www.wellcertified.com/certification/v1/standard/ (accessed on 16 July 2021).

52. Larsen, T.S.; Rohde, L.; Jønsson, K.T.; Rasmussen, B.; Jensen, R.L.; Knudsen, H.N.; Witterseh, T.; Bekö, G. IEQ-Compass—A tool for holistic evaluation of potential indoor environmental quality. *Build. Environ.* **2020**, *172*. [CrossRef]

53. CEN, EN 16798-1:2019 Energy Performance of Buildings—Ventilation for Buildings—Part 1: Indoor Environmental Input Parameters for Design and Assessment of Energy Performance of Buildings Addressing Indoor Air Quality, Thermal Environment, Lighting and Acous. 2019.

54. CEN, EN 16798-2:2019 Energy Performance of Buildings—Ventilation for Buildings—Part 2: Interpretation of the Requirements in EN 16798-1—Indoor Environmental Input Parameters for Design and Assessment of Energy Performance of Buildings Addressing Indoor Air Quality, Thermal Environment, Lighting and Acoustics (Module M1-6). 2019.

55. *Ventilation for Acceptable Indoor Air Quality, (ASHRAE Standard)*; R. American Society of Heating and and Air-Conditioning Engineers (ASHRAE): New York, NY, USA, 2004; Volume 2004, p. 53.

56. Dorizas, P.V.; de Groote, M.; Volt, J. *The Inner Value of a Building: Linking Indoor Environmental Quality and Energy Performance in Building Regulation*; Buildings Performance Institute Europe: Brussels, Belgium, 2018.

57. VELUX, Daylight Calculations and Measurements. Available online: https://www.velux.com/what-we-do/research-and-knowledge/deic-basic-book/daylight/daylight-calculations-and-measurements (accessed on 16 July 2021).

58. BPIE, How to Integrate Indoor Environmental Quality Long-Term Renovation Strategies. Available online: https://www.bpie.eu/publication/briefing-how-to-integrate-indoor-environmental-quality-within-national-long-term-renovation-strategies/ (accessed on 16 July 2021).

59. CISBE. *Environmental Design: Guide A*; CIBSE: London, UK, 2006.

60. Colantonio, A. Social sustainability: A review and critique of traditional versus emerging themes and assessment methods. In Proceedings of the 2nd International Conference on Whole Life Urban Sustainability and its Assessment, Loughborough, UK, 22–24 April 2009.

61. Polese, M.; Stren, R. *The Social Sustainability of Cities*; University of Toronto Press: Toronto, ON, Canada, 2000.

62. Bukovszki, V.; Balázs, R.; Mafé, C.; Reith, A. Six lessons learned by considering social sustainability in plus-energy neighbourhoods. In Proceedings of the 18th Vienna Congress on Sustainable Building, Vienna, Austria, 24–25 March 2021; pp. 26–28.

63. Bukovszki, V.; Magyari, Á.; Braun, M.K.; Párdi, K.; Reith, A. Energy modelling as a trigger for energy communities: A joint socio-technical perspective. *Energies* **2020**, *13*, 2274. [CrossRef]

64. Dempsey, N.; Bramley, G.; Power, S.; Brown, C. The social dimension of sustainable development: Defining urban social sustainability. *Sustain. Dev.* **2011**, *19*, 289–300. [CrossRef]

65. Takagi, H. *Interactive Evolutionary Computation: System Optimization Based on Human Subjective Evaluation*; IEEE: Piscataway, NJ, USA, 1998.

66. Ansell, C.K.; Torfing, J. *Public Innovation through Collaboration and Design*; Routledge: Abingdon, UK, 2014.

67. National Affordable Homes Agency. 721 Housing Quality Indicators (HQI) Form. 2007. Available online: https://assets.publishing.service.gov.uk/government/uploads/system/uploads/attachment_data/file/366634/721_hqi_form_4_apr_08_update_20080820153028.pdf (accessed on 16 July 2021).

68. Soja, E.W. The city and spatial justice. *Spat. Justice* **2009**, *1*, 1–5.

69. Thema, J.; Vondung, F. *EPOV Indicator Dashboard: Methodology Guidebook*; Wuppertal Institut für Klima, Umwelt, Energie GmbH: Wuppertal, Germany, 2020.

70. European Commission. The European Green Deal Investment Plan and Just Transition Mechanism Explained. 2020. Available online: https://ec.europa.eu/commission/presscorner/detail/en/qanda_20_24 (accessed on 16 July 2021).

71. Leidelmeijer, K.; Marlet, G.; Ponds, R.; Schulenberg, R.; van Woerkens, C.; van Ham, M.V.M. *Leefbaarometer 2.0: Instrumentontwikkeling*; Research en Advies: Amsterdam, The Netherlands, 2015.

72. City of Vancouver. Social Indicators and Trends: City of Vancouver Profile 2019. Vancouver, BC, Canada. 2019. Available online: https://vancouver.ca/files/cov/2020-01-22-social-indicators-and-trends-2019-city-demographic-profile.pdf (accessed on 16 July 2021).

73. European Commission. Sustainable Cities and Communities. In *Sustainable Development Indicators*; Eurostat: Luxembourg, 2020.

74. BREEAM. *BREEAM Communities Technical Manual*; BRE: Watford, UK, 2012.

75. European Commission. Glossary—Regional Policy. 2020. Available online: https://ec.europa.eu/regional_policy/en/policy/what/glossary/ (accessed on 16 July 2021).

76. ISO. *ISO 21542:2011(en), Building Construction—Accessibility and Usability of the Built Environment*; BREEAM Communities Technical Manual; BRE: Watford, UK, 2012.

77. Michalos, E. *Encyclopedia of Quality of Life and Well-Being Research*; Springer: Dordrecht, The Netherlands, 2014.

78. Scanlon Foundation. Mapping Social Cohesion. 2019. Available online: https://scanlonfoundation.org.au/wp-content/uploads/2019/11/Mapping-Social-Cohesion-2019-FINAL-3.pdf (accessed on 16 July 2021).

79. Anne, T.; Bass, R.M. Assessing Your Innovation District: A How-to Guide. 2018. Available online: https://www.brookings.edu/wp-content/uploads/2018/02/audit-handbook.pdf (accessed on 16 July 2021).

80. Paul, J.; Modi, A.; Patel, J. Predicting green product consumption using theory of planned behavior and reasoned action. *J. Retail. Consum. Serv.* **2016**, *29*, 123–134. [CrossRef]

81. van Trijp, H.C.M. *Encouraging Sustainable Behavior: Psychology and the Environment*; Taylor and Francis: Abingdon, UK, 2013.

82. Junker, R.G.; Azar, A.G.; Lopes, R.A.; Lindberg, K.B.; Reynders, G.; Relan, R.; Madsen, H. Characterizing the energy flexibility of buildings and districts. *Appl. Energy* **2018**, *225*. [CrossRef]

83. Verbeke, S.; Aerts, D.; Reynders, G.; Ma, Y.; Waide, P. (WSEE), Final Report on the Technical Support to the Development of a Smart Readiness Indicator for Buildings. 2020. Available online: https://op.europa.eu/en/publication-detail/-/publication/f9e6d89d-fbb1-11ea-b44f-01aa75ed71a1/language-en?WT.mc_id=Searchresult&WT.ria_c=37085&WT.ria_f=3608&WT.ria_ev=search (accessed on 14 June 2021).

84. Verbeke, S.; Aerts, D.; Rynders, G.; Ma, Y.; Waide, P. Summary of State of Affairs in 2nd Technical Support Study on the Smart Readiness Indicator for Buildings. 2020. Available online: https://smartreadinessindicator.eu/sites/smartreadinessindicator.eu/files/sri_summary_2nd_interim_report.pdf (accessed on 16 July 2021).

85. European Commission. Smart' Buildings—Smart Readiness Indicator (Arrangements for Rollout of Scheme). Available online: https://ec.europa.eu/info/law/better-regulation/have-your-say/initiatives/12365-Implementation-modalities-of-the-smart-readiness-indicator-for-buildings (accessed on 16 July 2021).

*Article*

# Optimized Energy and Air Quality Management of Shared Smart Buildings in the COVID-19 Scenario

Giuseppe Anastasi [1], Carlo Bartoli [2], Paolo Conti [2], Emanuele Crisostomi [2], Alessandro Franco [2,*], Sergio Saponara [1], Daniele Testi [2], Dimitri Thomopulos [2] and Carlo Vallati [1]

[1]  Department of Information Engineering (DII), University of Pisa, Largo Lucio Lazzarino, 56122 Pisa, Italy; giuseppe.anastasi@unipi.it (G.A.); sergio.saponara@unipi.it (S.S.); carlo.vallati@unipi.it (C.V.)

[2]  Department of Energy, Systems, Territory, and Constructions Engineering (DESTEC), University of Pisa, Largo Lucio Lazzarino, 56122 Pisa, Italy; carlo.bartoli@unipi.it (C.B.); paolo.conti@unipi.it (P.C.); emanuele.crisostomi@unipi.it (E.C.); daniele.testi@unipi.it (D.T.); dimitri.thomopulos@unipi.it (D.T.)

*  Correspondence: alessandro.franco@ing.unipi.it

**Abstract:** Worldwide increasing awareness of energy sustainability issues has been the main driver in developing the concepts of (Nearly) Zero Energy Buildings, where the reduced energy consumptions are (nearly) fully covered by power locally generated by renewable sources. At the same time, recent advances in Internet of Things technologies are among the main enablers of Smart Homes and Buildings. The transition of conventional buildings into active environments that process, elaborate and react to online measured environmental quantities is being accelerated by the aspects related to COVID-19, most notably in terms of air exchange and the monitoring of the density of occupants. In this paper, we address the problem of maximizing the energy efficiency and comfort perceived by occupants, defined in terms of thermal comfort, visual comfort and air quality. The case study of the University of Pisa is considered as a practical example to show preliminary results of the aggregation of environmental data.

**Keywords:** building dynamics; occupants' comfort; energy efficiency; information and communication technologies; COVID-19 scenario; human interaction

**Citation:** Anastasi, G.; Bartoli, C.; Conti, P.; Crisostomi, E.; Franco, A.; Saponara, S.; Testi, D.; Thomopulos, D.; Vallati, C. Optimized Energy and Air Quality Management of Shared Smart Buildings in the COVID-19 Scenario. *Energies* **2021**, *14*, 2124. https://doi.org/10.3390/en14082124

Academic Editors: Francesco Nocera, Anastasios Dounis and Jae-Weon Jeong

Received: 22 February 2021
Accepted: 8 April 2021
Published: 10 April 2021

**Publisher's Note:** MDPI stays neutral with regard to jurisdictional claims in published maps and institutional affiliations.

## 1. Introduction

Energy efficiency, indoor air quality and user comfort in buildings have received increasing attention in recent years, as they permit a reduction in the consumption of conventional fuels and greenhouse gas emissions, which are fundamental targets in sustainability programs. In addition, they contribute to reducing energy costs and improving the health conditions of building owners [1]. People can be considered as part of any building energy system as, according to the available literature, they spend about 80% of the time indoors—up to 90% in some European countries [2]. Buildings in general, and public buildings as well, are among the major energy consumers in cities, and again this is especially true in European countries. Considering the correlation among energy consumption, its environmental footprint and the implications for global warming [3], energy sustainability emerges as one of the key challenges for decision-makers and society in general. The recent COVID-19 pandemic situation has even further exacerbated the importance of the topic of energy use in public buildings.

A significant amount of energy is necessary in order to maintain the comfort level for the occupants through the operation of various appliances. Occupants can be considered responsible for energy use in buildings; for this motivation "psychology of energy saving" and strategies for energy efficiency have started receiving attention since the 1970s [4,5].

The building energy management and possible trade-offs between comfort and energy consumption have been the subject of several recent studies, such as [6,7]. At the same time, recent advances in Internet of Things (IoT) technologies are among the main enablers of Smart Homes (SHs) and Buildings (SBs). Roughly speaking, a SH or a SB may be defined as a highly automatized environment, where data collected by sensors are gathered and processed in an unsupervised fashion to affect all existing appliances and functionalities and improve the comfort of the people who spend their time in that environment [8–10].

While general awareness of the control of environmental variables has increased in the last few years, this interest has been recently increased by the spreading of the COVID-19 pandemic, particularly relevant for commercial and public buildings, such as those for educational purposes [11,12]. All countries have started planning post-lockdown activities and there is a growing concern regarding how social distancing measures and strict indoor air quality control can be enforced in shared buildings to prevent possible airborne virus transmission in indoor spaces.

In our vision, people should be able to enter SBs and learn about some basic environmental variables of the building, such as indoor temperature, air quality (e.g., in terms of $CO_2$ levels), and visual comfort, and they should be able to interact with the energy systems (e.g., the HVAC—Heating, Ventilation and Air Conditioning—system and lighting systems) up to a certain allowed extent (e.g., for security reasons), combined with natural resources (day lighting, outside temperature, etc.).

From this perspective, people may be willing to know the values of indoor environmental variables and be allowed to control HVAC actuators to improve some of such quantities according to their perceived comfort level. In this regard, people may be seen as mobile sensors that provide further indications to the Building Management System (BMS), and the BMS should be permitted to directly interact with occupants, receiving their feedback in different ways. In general, an elevated performance of the HVAC system can be achieved by radically improving the control performance, which is an important issue in HVAC systems [13]. Ventilation devices, as an important subsystem of the HVAC system, are operated to maintain an appropriate Indoor Air Quality (IAQ). Ventilation systems contribute to about 25–30% of HVAC energy use [14].

A trade-off problem exists in utilizing the several components that can ensure a desired IAQ with minimal energy consumption and reduced energy costs. Currently, such a problem is often solved by adopting a centralized optimization approach. In the recent literature, various methods have been proposed to address the optimization problem of the energy management of buildings and, in general, the main goal of each proposal is to find a balance between user comfort and energy consumption [15–18].

However, efficient building energy consumption and the maintenance of a high level of comfort are still challenging tasks: various heterogeneous variables and parameters affect the problem. In general, the proposed framework of balancing energy consumption and occupants' preferences by adapting building controls to users' activities and requirements does not consider the direct interaction of the users.

Accordingly, in this work we plan to revisit current energy/environment management strategies of smart buildings by including active occupants who interact with the SB through their smartphones. They will act as both sensors and actuators to complement control systems already installed in buildings (e.g., surveillance cameras to be used in the COVID-19 era to detect and count people occupying a room, as well as to measure if social distancing is maintained). The optimized real-time HVAC control strategy should be found through a multi-faceted optimization problem that considers the different and possibly contrasting desires of people, together with energy consumption reduction in public buildings. The present work tries to include the aforementioned "new applications" within classic BMS optimization problems, and tries to address the challenges of combining classic HVAC or lighting control problems with the new problems arising in a COVID-19 context. As usual, a new challenge (here, the COVID-19 pandemic) may also give rise to new ideas, and in our opinion the COVID-19 pandemic may be a "killer application" to eventually realize truly smart buildings, where users interact with the environmental variables.

In this work, some experimental activities carried out in the classrooms of the University of Pisa will be used to support the proposed methodologies. We shall leverage on available pre-pandemic historical data to show the feasibility of our proposal.

Hereafter, the paper is organized as follows: after the introduction, Section 2 presents the proposed methodology. Section 3 shows an overview of sensors used for monitoring action and Section 4 presents the logic of the control system. Section 5 shows some experimental results obtained in pre-pandemic conditions in a building of the University of Pisa and Section 6 describes the implementation of the proposed approach to define feedback on the control system. Conclusions are drawn in Section 7.

## 2. ICT-Based Methodology for Balancing Energy Efficiency and Comfort

As mentioned in the introductory section, the problem of balancing the trade-off between minimum energy cost and maximum comfort has been widely explored in the literature. Different optimization methods have been proposed to tackle such a problem, but all of them heavily rely on real-time measured data, and in general on Information and Communication Technologies (ICTs) as the enabling technology to collect and analyze such data.

Data can be acquired through traditional sensors: Internet of Things (IoT) technologies are successfully utilized in real environments to make IoT-based smart buildings successful. In this work, however, we propose a different approach: further exploiting a direct interaction with the occupants by using the potential of their portable devices and communication systems.

Considering the abovementioned possibilities, this study proposes the use of special sensors and advanced ICTs for monitoring environmental conditions indoors, both with the purpose of controlling climatic conditions, somewhat achieving a desired comfort with reduced costs, but also with the further objective of preventing the diffusion of SARS-CoV-2 (and other future viruses with pandemic potential) to support the reopening of public and private buildings with a higher level of safety. To reduce the risk of infection by SARS-CoV-2 and other respiratory viruses, the main strategy is to control the probability of contacts. In fact, SARS-CoV-2 is transmitted among people through inhalation and exhalation. As the occupation density increases, the probability of virus transmission increases, so that the occupancy/density of the building becomes another reason to trigger air exchanges in addition to air quality control. On the other hand, the increased rate of air changes can be highly energy consuming, and the reduced possibility of using some typical measurements such as air recirculation introduces a further criticality because the power of the installed heat exchangers and thermal generators could be not high enough to match the demand of the building and guarantee an acceptable level of indoor thermal comfort.

Considering the above exposed objectives, we propose two main elements. The first is to integrate environmental sensors and tele/thermal cameras to evaluate environmental variables (e.g., temperature, humidity, $CO_2$ concentration), and, thanks to specific Artificial Intelligence (AI) algorithms, to check in real time the maintenance of safety measures, such as social distancing and the correct use of masks, and the possible presence of feverish people in closed spaces (e.g., schools and university classrooms, entrance queues to sanitary buildings, waiting rooms, public/private buildings in general).

The second relevant element proposed in the paper is the possibility of obtaining direct interaction with the occupants to check the main environmental variables and comfort conditions. The occupants' interaction can be achieved by using personal devices (e.g., smartphones and tablets) by means of specific human–computer interfaces designed for improved user experience.

The control system must acquire the inputs from the occupants, as well as acquire the data by the various physical sensors deployed in the building, to find the best trade-off between normative requirements, perceived comfort, energy use and safety.

The "sensorization" of critical environments and the use of AI algorithms for data analysis offer numerous advantages. Firstly, this does not require the presence of a person in charge of monitoring the accesses and measuring the facial temperature of all users, which is critical in buildings with multiple access points (in addition, facial temperature can evolve over the course of a day). Consequently, no personnel are required to check the correct use of safety devices or keep physical distancing among people. Secondly, room air exchanges can be adjusted according to actual needs (e.g., based on $CO_2$ concentration, temperature and humidity conditions). The monitoring of these variables will result in unsupervised automatic actions (e.g., automatic adaptation of air changes in the environment) and/or supervised ones (e.g., the reporting of non-virtuous behaviors or the presence of feverish individuals), and integration into existing organizational structures of new methodologies for passive and active COVID-19 monitoring.

## 3. Sensors for Monitoring and Available Data

Strategies for the sustainable design and management of shared buildings should also promote healthy and comfortable indoor environments. Energy saving potential and control of comfort conditions are surely aided by the measurement of specific variables by means of specific sensors. Occupancy sensors, for example, have a potential to significantly reduce energy use by switching off electric loads when an area is vacated or when its occupation is highly reduced.

Regarding the comfort parameters, it is important to control the values of temperature and relative humidity (RH) and limit the concentration of pollutants, such as $CO_2$. Nowadays, after the COVID-19 pandemic experience, with the accurate control of the indoor environment it will be easier to prevent virus transmission. It is recognized that most of the COVID-19 infections happen in public spaces, so that an accurate check of the occupant distribution pattern can reduce the infection rate.

It is well known that a series of parameters have to be for the risk transmission in indoor spaces, and those for a given value of area, height and volume are surely connected with ventilation (mechanical or natural), air recirculation and air flows, humidity and layout and use of the spaces (classroom, corridors, bar, multi-functional spaces). Therefore, accurate control requires first of all the knowledge of indoor air quality parameters, such as temperature, relative humidity and carbon dioxide concentration, as well as an accurate estimation of indoor occupation.

### 3.1. Sensors for $CO_2$ Concentration, Temperature, and Humidity Detection

To accurately check air quality parameters, sensors for the simultaneous measurement of the different indoor variables can be used. A lot of sensors are available on the market that can be placed indoors. An interesting example could be Chauvin Arnoux, in particular, model C.A 1510. The C.A 1510 is an instrument for measuring physical quantities that provides measurement of:

- Carbon dioxide ($CO_2$) concentration in air;
- Internal temperature (T);
- Relative humidity (RH).

The characteristics of the sensors are the following ones:

- $CO_2$ concentration is measured with infrared technology. The measurement range is from 0 to 5000 ppm; the intrinsic uncertainty is of the order of $\pm 3\%$ ($\pm 50$ ppm) at 25 °C and 1 bar of pressure. The instrument has a resolution of 1 ppm and it can operate at temperatures in the range between $-10$ and 45 °C;
- The measurement of temperature (T) is obtained by means of a CMOS sensor that can provide a relatively accurate value, with an uncertainty of $\pm 0.5$ °C in the range between $-10$ and 60 °C;
- The measurement of relative humidity (RH) is obtained by means of a capacitive sensor that permits the acquirement of values from 5% to 95% RH. The uncertainty is $\pm 2\%$ in the range from 10% to 90% RH and $\pm 3\%$ RH outside this range. Moreover, the instrument has a resolution of 0.1% RH and a hysteresis of 1% RH.

Figure 1 provides a snapshot of the instrument, in which the instantaneous values of $CO_2$ concentration (in ppm), temperature (in °C) and relative humidity (in %) are reported on the screen, and the typical results of an experimental analysis concerning $CO_2$ concentration and temperature evolution during an indoor measurement are also reported.

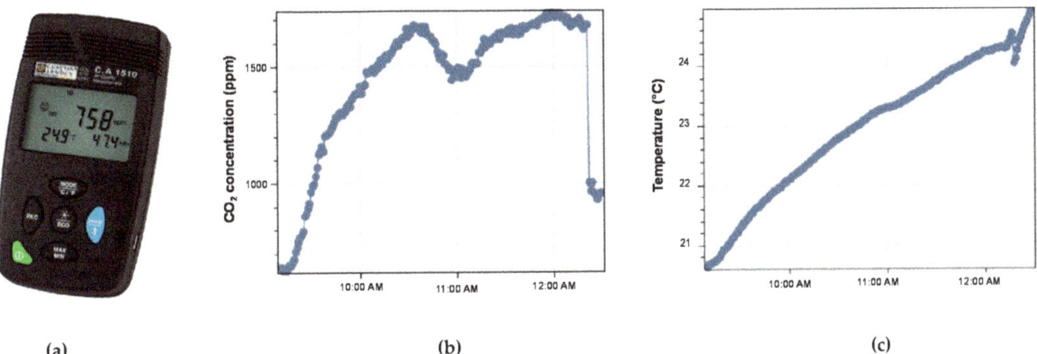

(a)    (b)    (c)

**Figure 1.** A snapshot of CA 1510 instrument (**a**) and typical data acquired for $CO_2$ concentration (**b**) and temperature (**c**).

### 3.2. Sensors for Object Detection and Control of the Presence

In recent times, the detection of objects in an image can be simply solved thanks to the improvements in computer vision and deep learning. Object detection systems are based on the concept of placing a bounding box around the objects and associating the correct object's category with each bounding box. Deep Learning (DL) is an effective method to perform object detection, and in fact it is increasingly applied to problems of social distancing [19]. DL, for instance, can be applied to detection through bounding box information with approximate models as in [20], or exploited with hybrid models of Computer Vision and Deep Neural Networks (DNNs) for an automated detection as in [21], or Convolutional Neural Network (CNN) as in [22]. In particular, the model proposed by a research group of the University of Pisa for smart city applications [22,23] consists of four stages. The first level is represented by the introduction of the images into the input layer, then the regional proposals are extracted, after that the features are computed by CNN, and finally these features are classified, as exposed in Figure 2. In particular, Region Based Convolutional Neural Network (R-CNN) uses selective search algorithms in order to define region proposals.

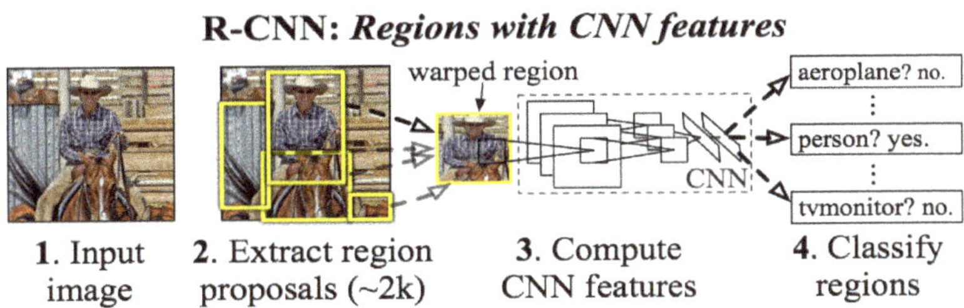


**Figure 2.** Concept for Region Based Convolutional Neural Network (R-CNN) detector: a schematic diagram.

The proposed detection approach can also be applied to images acquired by thermal cameras to establish a complete AI system for people tracking, social distancing control and facial temperature monitoring without direct interaction with the occupants. By means of the elaboration of the images acquired, measured distance (D) between the center of each bounding box for a detected person can be defined according to Figure 3a. This figure also reports an example of a social distancing check (the green box corresponds to a safe situation, while the red box indicates an unsafe situation).

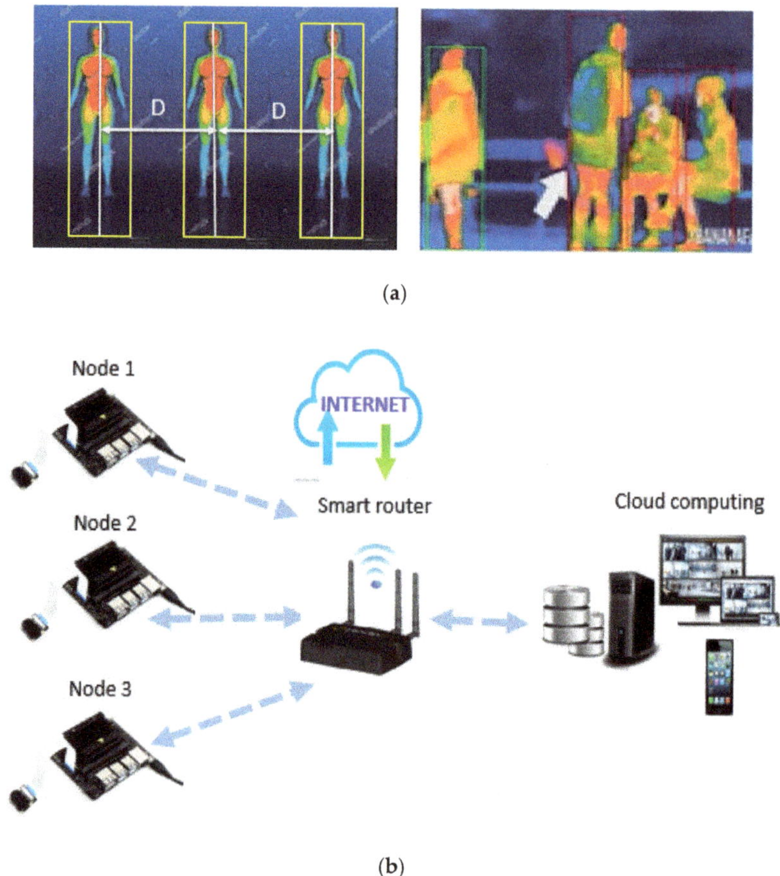



**Figure 3.** The method for checking the distancing (**a**) and the concept of smart surveillance distributed video system (**b**).

An application in indoor and outdoor scenarios for COVID-19 people detection and social distance checking has been proposed in [22]. It is to be noted that the proposed algorithm has been implemented in real-time on a low-cost embedded platform, such as Jetson Nano. Moreover, as shown in Figure 3b, the system can be scaled from a single node to a solution with multiple distributed nodes, thus scaling the application area from a single building to a district and then to a smart city.

## 4. A User-Centric Control System

As we have already discussed in the introduction, the main objective of the study referred in the manuscript is to propose systems where users can influence the functioning of climate automatic control systems, with the further objective of solving an optimization problem with contrasting objectives, aiming at obtaining a convenient trade-off between comfort and energy consumption. The main functions of a building management control system are sensing the environmental factors by measurements and optimizing control strategies based on the current and predictive states of the building and occupancy [24].

Similar ideas have already been widely explored and applied in other contexts, see, for instance, building automation applications in hotels or cruise ships, where guests may be permitted to both interact with the climate/lighting systems in their own rooms and also in shared spaces (e.g., receptions or lounge areas) or in public buildings with reference to lighting control systems [25]. Besides, similar opportunities are not usually available in residential/commercial buildings, such as shopping centers, universities or restaurants, which are the scope of this paper.

Following typical trends in similar applications, we assume that interactions between a building management control system and its occupants should be based on suitably developed smartphone/tablet apps. By subscribing to such apps, a guest accepts to share his/her personal information (e.g., locally measured environmental data, occupancy data, data regarding the usage of the building) to gain access to the building management control system. From this perspective, building guests/customers serve as mobile sensors which enrich the already existing sensor network. Customers may share information directly measured by their smartphones (e.g., temperature, humidity, distance from other customers, density of occupancy, mobility patterns inside the facility) and also communicate textual information (e.g., a customer may inform the system that he/she is feeling cold). The data acquired by the sensors (of temperature, concentration of specific pollutants, humidity, and illuminance) are thus integrated with information collected from app subscribers. Such data may be used for safety/security purposes (e.g., occupancy data), but also to solve the complex problems of balancing energy consumption and comfort [26].

Accordingly, the control platform aggregates all the data and tries to solve a multi-objective optimization problem for finding the most convenient trade-off between energy efficiency and comfort. Then, it will send the appropriate control signals to the available actuators (i.e., the HVAC system). Additionally, interaction with acoustic, environmental, and lighting systems, with the purpose of obtaining occupant-centered acoustic and lighting control, should be considered as well. Overall, the system behaves as a cyber-physical system where individuals interact with the building management system.

In specific situations, in addition to behaving as mobile sensors, occupants may also be allowed to perform specific actuating actions. This may be orchestrated by assigning appropriate priorities to single occupants (e.g., teachers, technic staff, and students in a university framework) and by allowing some categories of occupants to take specific control actions (e.g., changing the temperature set-point, changing the frequency of air exchangers, or interacting with the blind or lighting system). Within the platform, all non-main commercial building electric loads, Miscellaneous Electric Loads (MELs), which are not controlled by the energy management system, will also take an important role, as they contribute a significant portion of the energy consumption.

To implement and fully exploit the aforementioned capabilities and obtain the expected results, the ICT platform should have the following features:

- Most currently existing building control and management systems rely on predictive models and the simulation of occupancies. However, the accuracy of predictions of actual occupancies is obviously questionable and subject to unexpected anomalous deviations. Conversely, the proposed ICT platform will not require buildings and occupancy models, but will rely on data acquired in real-time by appropriately installed static sensors, and integrated with dynamic information. By embedding learning and self-adapting capabilities, the platform will be permitted to define a "just-enough-accurate" model for the building/occupants' behaviors, which can be used for optimal decisions.
- In order to prevent the problems connected to the occupants that can override the decisions of the ICT platform and/or not adopt its recommendations, the control system will be able to define optimal strategies for the operating systems, like HVAC. To this aim, an iterative procedure based on the learning and self-adapting mechanisms embedded in the control system is required. Such a procedure will provide the control strategy, will evaluate how much the occupants' actions are consistent with

these decisions, and will calibrate the decision-making mechanism to "close the gap" between decisions and occupants' actions.

- Energy awareness can be further boosted by the direct interaction between occupants and the platform through a kind of gamification technique in which wise behaviors and the resulting greater energy efficiency will be consequent to the participation in this "social game".
- The interaction between the ICT platform and occupants will be designed with the objective of maximizing the energy-saving with the constraints of maintaining a comfortable working environment.

## 5. Building Management Control as an Optimization Problem

The considered building optimization strategy assumes that prior knowledge of the characteristics of the building is given, and occupancy data and external climatic conditions are either measured or estimated. In particular, data acquired from the sensors and from the occupants allow the Building Management System (BMS) to evaluate the energy consumption due to both the lighting system and the HVAC operation (for temperature and ventilation control), the indoor comfort conditions, and to directly interact with the various active systems.

The proposed BMS and its feedback control scheme may be summarized in Figure 4. The environmental data of the building are collected by sensors, enriched with the sensors of occupants who have installed the building application (e.g., sensors available in personal smartphones).

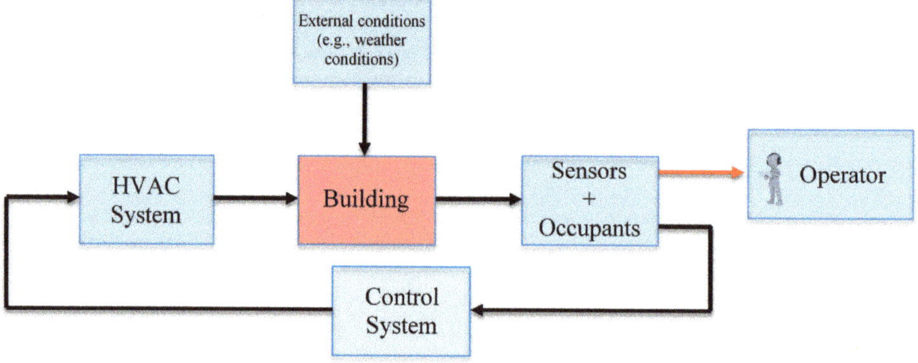


**Figure 4.** Control scheme for building control.

The flow of critical data, represented in red in Figure 4, is directly communicated to the operator, who is in charge of taking actions to respect the safety protocol. This includes the counting of the occupants of a building/room, who should not exceed a safety threshold (e.g., depending on the size of the room), the maintenance of safe distances (e.g., below 1 m) among occupants, and the temperature of single individuals as measured by the thermal cameras (e.g., below 37.5 °C).

If any of the aforementioned rules is infringed, then ad hoc measures should be taken. Conversely, other non-critical environmental quantities are directly elaborated by the control system, possibly changing the set-points of the HVAC system. While the full list of controlled environmental quantities depends on the specific interests, or on the available sensors, they include quantities such as temperature, air quality (e.g., level of $CO_2$), moisture, lighting, and level of noise. Similarly, while only the HVAC system is reported in Figure 4, in general, other available actuators should include the lighting system as well. The lighting system could allow for simple ON/OFF choices, or could dim the quantity of provided light, or even provide more sophisticated levels of control, e.g., if the light can change the color, change the duration of colors, or associate the light control with music.

The control system may be implemented as an intelligent system that computes the optimal control actions as the outcomes of an optimization problem, which can be formulated as a kind of bounded minimization problem:

$$\begin{cases} \min f(x) \\ \underline{x} \leq x \leq \overline{x} \end{cases} \tag{1}$$

where $x$ is a vector that contains all the monitored environmental quantities (e.g., most notably, temperature, concentration of $CO_2$, lighting, moisture). In Equation (1), $\underline{x}$ and $\overline{x}$ denote the vectors of lower and upper bounds, respectively, of the environmental quantities (e.g., in the case of temperature control, one may constrain the temperature to lie within $17\,^{\circ}C$ and $25\,^{\circ}C$; or in the case of $CO_2$ concentration, a classic upper bound of 1500 ppm can be defined, as it is discussed in Section 6 of the paper).

The function $f(x)$ represents some cost function of interest that one aims at minimizing to optimize the utility of occupants [27]. In our work, we assume that this cost function consists of two terms, as follows:

$$f(x) = w_{discomfort}\, f_1(x) + w_{energy}\, f_2(x) \tag{2}$$

where $f_1(x)$ represents the cost associated with discomfort, and $f_2(x)$ is the energy cost (which we shall compute in terms of energy consumption but could be alternatively translated into the associated cost of energy).

In Equation (2), $w_{discomfort}$ and $w_{energy}$ are two coefficients that can be used to either prioritize comfort or price, or, in general, a convenient trade-off between the two components that are combined in the overall cost function. In addition, such weights should also be used to normalize the different quantities that are combined in the cost function, so that each component has, on average, the same impact on the overall objective function.

After the normalization step, it is possible to think of the two objective functions as expressed in a dimensionless form. We now discuss how it is possible to normalize the two functions. Concerning the component of the objective function related to energy, the total value of used energy, expressed in kWh, can be used as a normalization factor, and the objective function can be defined in a dimensionless way considering the distance from such a reference value.

Energy consumption can be expressed as the sum of the energy consumption determined by the individual components of the building management system in a given period, as:

$$E = E_{Temp} + E_{RH} + E_{IAQ} + E_{lux}, \tag{3}$$

where the four terms appearing in Equation (3) denote the energy required to accomplish the regulation of the indoor temperature, relative humidity, indoor air quality, and lighting. Energy consumption and its components are functions of the vector of variables $x$, e.g., $E(x)$. However, for ease of notation, they are expressed simply as single variables, e.g., $E$. If we let $E_{Ref}$ denote a reference value for energy consumption without any smart control action activated, then the consumed energy can be expressed in a dimensionless form as:

$$f_2(x) = \frac{E}{E_{Ref}} \tag{4}$$

Conversely, the definition of a specific objective function for comfort is not trivial. In fact, the concept of comfort cannot be easily translated into a quantitative indicator. Usually, thermal comfort is indicated through a temperature index, so that an optimal value can be defined and the operation of the HVAC system is used for maintaining comfortable indoor temperature and humidity values. Visual comfort is usually evaluated with the brilliance level. Air quality can be indicated by a $CO_2$ concentration index; both natural and mechanical ventilation systems are employed for maintaining an acceptable $CO_2$ concentration level in buildings.

In our work, we can adopt some dimensionless parameters similar to the Predicted Mean Vote (PMV) to evaluate the objective function for comfort. We propose some modified dimensionless indicators that can be defined according to Fanger's model, and may be implemented in a feedback loop, as in the spirit of Figure 4, considering thermal comfort level, humidity, IAQ level, and luminance [28,29].

According to the original idea and consistently with the minimization problem expressed in Equation (1), the comfort level is expressed in terms of a discomfort parameter, $D$, that one wishes to minimize, which can be defined in dimensionless terms as:

$$f_2(x) = D = \frac{|I - I_{set}|}{I_{set}} \tag{5}$$

$$I = I_{Temp} + I_{RH} + I_{Cco2} + I_{lux} \tag{6}$$

Considering the definition of the discomfort function $D$, $I$ is the value of the typical comfort indicator (of temperature, relative humidity, $CO_2$ concentration and luminance), while $I_{set}$ is the desired set-point of each comfort parameter. In this way, the value of 0 corresponds to the highest comfort level, while the maximum is not upper bounded. Similarly to energy consumption, discomfort and comfort indicators are also functions of the vector of variables $x$; however, for ease of notation, they are expressed simply as single variables.

The optimization problem defined by Equation (1) can be solved in a convenient way using classic convex optimization tools, if the cost function $f(x)$ is computed by adding single functions that are convex with respect to the single environmental quantities and have their minimum in the preferred set-point (e.g., 21 °C in the case of the indoor temperature). Accordingly, if one prioritizes comfort aspects (by setting $w_{energy}$ equal to zero), then the Control System works to guarantee some specific parameters, such as, for example, a constant temperature level of 21 °C (during the winter season), using simple Proportional Integral (PI) control rules. Conversely, if one prioritizes price aspects, then the temperature will be around the minimum allowed value in winter time, and around the maximum allowed value in summer time, to minimize the cost of the HVAC systems.

## 6. Preliminary Analysis and Data Acquired in a Pre-Pandemic Scenario: Identification of Meaningful Variables

An educational building has been chosen as the preliminary target setting for this study, for establishing a connection between occupancy and some specific physical parameters such as temperature, relative humidity, and $CO_2$ concentration. Universities are part of the tertiary sector and are an ideal example of shared buildings, as they include areas that are only shared among few authorized individuals (e.g., departments and teachers' offices) and other parts shared with hundreds/thousands of students (e.g., teaching rooms and laboratories). In addition, universities represent an ideal case study as they are also characterized by relatively high values of energy consumption, which could be easily reduced by applying basic building automation practices. They obviously lend themselves to retrofitting both in terms of structure and behavior [30]. Finally, since universities are educational institutions, they also have the potential to educate students to adopt virtuous behaviors and have an influential impact on society, and should be obliged to act in a sustainable manner [31].

An extensive pilot study aimed at obtaining a preliminary calibration of the method has been carried out in several engineering classrooms of the University of Pisa, where the devices described in Section 3.1 have been used to measure the correlation of the air quality parameters in terms of temperature, relative humidity and $CO_2$ concentration, and how they vary with different values of occupancy. In principle, the connection between such physical parameters is well known in the literature, but in the case of temperature and RH, it is difficult to establish a quantitative correlation. This appears to be possible considering

$CO_2$ concentration, as shown in some original studies available in the literature, such as [32–34].

In particular, the building involved in the experimental campaign had been originally designed for commercial activities and it has been refurbished for educational purposes during 2006. The building contains nine classrooms of different sizes, and all of them but one are characterized by high ceilings (more than 5 m high). Additionally, most of the rooms are equipped with blinds to shade the lights during the day, which are particularly convenient when teachers project slides during their lessons.

All classrooms are characterized by a large penetration of light, but thanks to the high ceilings, they are also characterized by a large volume, and by a large volume-to-surface ratio (sensibly larger than 5). In this structure, six different classrooms, characterized by different sizes, volumes, and capacity, have been selected for the monitoring of the indoor air quality, and the main characteristics are reported in Table 1. Considering the typical form of the various classrooms, we considered the average value of the measurements of two or three different sensors, disposed according to the schemes provided in Figure 5a, in the case of two sensors, and Figure 5b, in the case of three sensors placed inside the classroom. Another sensor is placed outside the classroom to collect the reference level. An example is provided in Figure 5b (see the yellow point).

**Table 1.** Reference data for the 6 monitored classrooms in the selected educational building.

| Room | Maximum Number of Occupants ($n$) | Floor Surface (m²) | Volume (m³) | Ratio Vol./Surface (m³/m²) | Vol. Per Student at Full Occupation (m³) |
|---|---|---|---|---|---|
| 1 | 309 | 286 | 1587 | 5.55 | 5.20 |
| 2 | 208 | 216 | 1220 | 5.65 | 5.75 |
| 3 | 109 | 130 | 721 | 5.54 | 6.61 |
| 4 | 196 | 197 | 1093 | 5.54 | 5.52 |
| 5 | 104 | 129 | 717 | 5.55 | 6.88 |
| 6 | 140 | 131 | 439 | 3.34 | 3.12 |

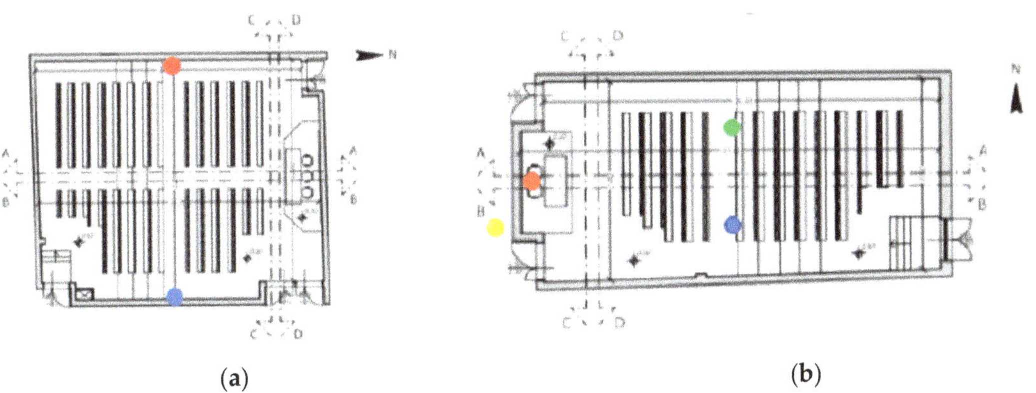

(a)       (b)

**Figure 5.** The position of the sensors in the two classrooms of different sizes: case of two sensors (**a**) and case of three sensors (**b**).

The analysis of the data acquired during the monitoring campaign in the different classrooms shows that $CO_2$ concentration quickly increases in time, with a rate that depends on the number of occupants, as shown, for instance, in Figure 5 for one of the monitored classrooms.

The data measured during the experimental analysis exhibit a clear correlation with the number of individuals inside the room and with other general indicators of the size of the room, such as the average volume available for each occupant. Measurements have been obtained during actual real operating conditions (during lessons or during periods of examination). To establish a direct correlation with the occupancy, ventilation was not disabled until a concentration rate of 1500 ppm had been observed. At this point, ventilation was activated again, as a concentration above 1500 ppm should be avoided [35,36] to maintain healthy indoor conditions, as suggested by the main Technical Standards for the analysis of indoor spaces.

As previously stated, during the first phase of the experimental analysis, when classes start the specific didactic activity (lesson or examination) and the mechanical ventilation is disabled, then it is possible to observe a quite linear correlation between the increase in $CO_2$ concentration rate and the specific volume per occupant.

A first-order fitting of the data in Table 2 leads to equations such as:

$$\frac{V}{n_{occ}} = \dot{r}\,\frac{1}{\frac{dC_{\{CO_2\}}(t)}{dt}} \tag{7}$$

where $\frac{V}{n_{occ}}$ is the net volume per number of occupants, $C_{\{CO_2\}}$ is the concentration of $CO_2$, and $\dot{r}$ is a proportional factor that can be interpreted as a kind of $CO_2$ production rate per person, typical of the occupants and of the specific activity carried out inside the room.

**Table 2.** Results of the experimental analysis.

| Room | Number of Occupants During Activity | Maximum Allowed Number of Occupants for the Classroom | Volume Available for Each Student (m$^3$) | $CO_2$ Concentration at the Beginning of Experience (ppm) | Time when $CO_2$ Threshold (1500 ppm) is Overcome (min) | $CO_2$ Variation up to the Threshold Value (ppm/s) | Duration of Monitoring Activity (min) |
|---|---|---|---|---|---|---|---|
| 1#1 | 58 | 305 | 27.36 | 678 | 95 | 0.144 | 131 |
| 1#2 | 93 | 305 | 17.06 | 596 | 78 | 0.193 | 192 |
| 2#1 | 168 | 212 | 7.26 | 1138 | 10 | 0.603 | 120 |
| 2#2 | 106 | 212 | 11.50 | 1200 | 21 | 0.238 | 140 |
| 3 | 32 | 148 | 22.53 | 725 | - | 0.191 | 53 |
| 4 | 146 | 198 | 7.48 | 791 | 24 | 0.492 | 105 |
| 5 | 54 | 104 | 13.27 | 1257 | 8 | 0.506 | 50 |
| 6 | 50 | 140 | 8.78 | 695 | 27 | 0.496 | 82 |

In this way, using an accurate estimate of $\dot{r}$, if the volume of the room is known beforehand, then, by measuring variations in $CO_2$ concentration over a given time, Equation (7) can be used to estimate the number of occupants in the classroom. Combining this piece of information with the one that is acquired through thermal cameras, safety protocols can be activated if the number of occupants exceeds the maximum number that is allowed. Besides, the level of $CO_2$ concentration can also be used to activate the ventilation system on demand, e.g., when a well-defined threshold value (for example, of 1500 ppm) is achieved, avoiding using ventilation when it is not required, to reduce energy consumption or, in general, activating the air ventilation in correlation with the estimated occupation of the classroom.

## 7. Critical Discussion and Future Research

The philosophy of the Control System exposed in Section 5 and summarized in Figure 4 may also be made more sophisticated and better performing if AI algorithms are used to learn the nominal behaviors of occupants. This implies that one could learn the typical occupancy patterns of the building, typical daily (or seasonal) variations of temperature and of other periodic environmental variables, so that the Control System may also predict future values of the environmental quantities and take pro-active actions in advance (e.g., switch on heating ahead of the arrival of students in the morning during winter days).

Although we have not explored this possibility yet, a further innovative step is to directly involve occupants/students, and in general of all the individuals who make use of the shared spaces, such as in shopping centers, public offices of commercial activities in general, to directly interact with the actuators that control the operation of HVAC systems. The risk of this implementation is that individuals with opposite desires give contrasting commands to the HVAC system (e.g., two different students in the same room may either want to switch on or off air conditioning). Accordingly, this possibility may be given only to specific authorized building users (e.g., teachers in our university case study) to prevent clashing situations from occurring.

In addition to the classic air quality control discussed so far, another important innovative aspect of our work is the utilization of AI algorithms and cameras operating in visible and infrared for greater accuracy in measurement (facial temperature), which will always be compensated by algorithms of dynamic calibration that consider environmental conditions (ambient temperature and humidity). The thermal imaging camera will also be combined with a camera in the visible range that can provide information about people's social behaviors (social distancing, correct use of masks, etc.) by applying artificial intelligence algorithms. More in detail, AI algorithms will also be used for "feature extraction" purposes, and a subsequent classification will be performed based on Deep Neural Networks (DNNs) to carry out the people detection and people counting phases (which is expected to be more accurate than the counting based on the ramp of $CO_2$ concentration). In particular, once individuals have been identified, they will be included in "bounding boxes", as described in Section 3.2, to detect the relative distance (determining if social distancing is respected) and to detect the face in order to determine whether the mouth and nose are appropriately covered with the mask, and also to determine the facial temperature in each bounding box. If safety issues are observed, this information will follow the red path shown in Figure 6, and an operator will be informed.

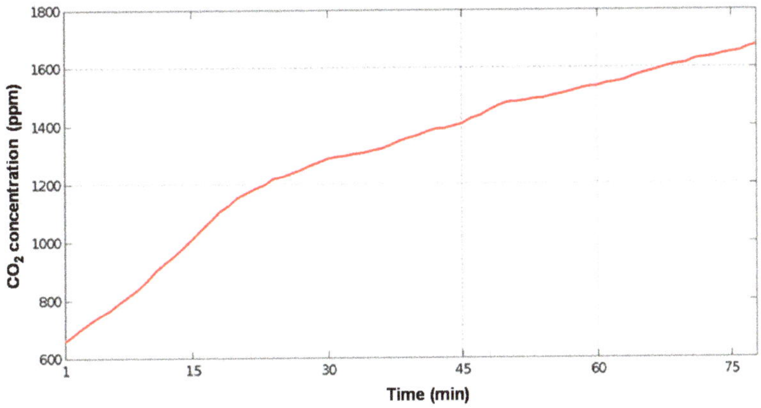

**Figure 6.** Results of typical experimental monitoring of $CO_2$ concentration in a classroom.

It is important to remark that while the main focus of our proposal is related to buildings, and indoor environments in general, the same philosophy we are applying here may be reused with little modifications in residential areas, and extended to include other sensors that we are not discussing here (e.g., measurement of biometric parameters).

More in general, what we are proposing is to consider human behaviors through the multi and interdisciplinary objective of building science, behavioral science, social science, data analysis, user experience based design, automation and control design. As human behavior is complex, it is challenging in general to address all the possible ways in which humans can actively interact with "buildings", or with the energy aspects of buildings in particular. Human factors can also be considered as an interesting driving source of innovation for energy efficiency in the built environment that contributes to achieving 2020 net-zero-energy buildings and 2050 post carbon goals set by the Paris Agreement [37].

Finally, the idea of active buildings with automated management of the indoor environment proposed in this paper can also be easily generalized to outdoor environments to contribute to the realization of smart cities. Again, in a smart city context, we envisage the utilization of individuals as mobile sensors that have the ability to interact with the available actuators (for instance, outdoor lamps may dim the provided lighting based on the presence of individuals, or cameras may be used to monitor the correct and safe utilization of public areas). More in general, such active monitoring paradigms should be used to support energy and urban planners in the creation of human-centered energy policies, programs, codes, and standards.

## 8. Conclusions

New approaches for designing and controlling the operation of HVAC systems are necessary and urgent to make indoor spaces safe and comfortable without compromising energy efficiency. The installation of instruments for indoor air quality monitoring is paramount, together with the implementation of building management and control systems.

This paper presents a methodology for the optimal control of safety, comfort and energy consumption parameters in smart buildings based on a more direct interaction with the occupants too. Knowledge of the occupancy profile, obtained directly by means of cameras or indirectly by means of measurements of air quality parameters, represents a step forward compared to current design and operation procedures suggested by technical standards. Interaction with the occupants through active participation is considered to be a relevant element of the methodology.

The argument specifically developed is that the classic topic of the optimization of a Building Management System (BMS) is becoming even more relevant in a COVID-19 context. Indeed, the functionalities of the same BMS should be further extended to include people counting, the monitoring and display of air quality and the need for air exchange (which plays a role in the spreading of the virus), and also for checking the temperature of individuals entering the buildings and enforcing that due distances are maintained. The optimization can be obtained by means of a cost function that combines different aspects, both in terms of energy consumption and of energy comfort, related to the operation of the HVAC systems, lighting systems, and the miscellaneous energy loads. The optimization method relies upon the definition of a certain number of constraints, such as, for example, the maximum occupancy of the building due to the COVID-19 pandemic and new indicators for comfort, considering the increased needs of air quality for human health. The application of the methodology to a test case represented by an educational building of the University of Pisa is proposed in this paper, evidencing the possible role of conventional sensors (such as temperature, humidity and $CO_2$ concentration) together with smart sensors.

The case proposed is just an example of a Shared Smart Building, and we tried to extract the main common features so that the same paradigm can also be applied to other examples of interest (this includes shopping centers, shared working facilities, sport and leisure centers, etc.). In most of these cases, the current practice is that people are in charge of measuring the temperature of customers or individuals at the entrance, people counting measures are not always strictly enforced, and proper distancing is not always guaranteed. Our vision is to make this automatic and embedded within the BMS.

Moreover, the method developed could also be extrapolated to external environments to contribute to the realization of smart cities and, in general, to support energy and urban planners in the creation of human-centered energy policies, programs, codes, and standards.

**Author Contributions:** Conceptualization, A.F., E.C., S.S.; methodology, A.F., E.C., S.S.; formal analysis, A.F., E.C., S.S.; data curation, A.F., E.C., S.S.; writing—original draft preparation, A.F., E.C., S.S.; writing—review and editing, A.F., E.C., S.S., G.A., C.B., P.C., D.T. (Daniele Testi), D.T. (Dimitri Thomopulos), C.V. All authors have read and agreed to the published version of the manuscript.

**Funding:** This research is possible thanks to the economic support of the University of Pisa for the participation in the EERA Consortium.

**Institutional Review Board Statement:** Not applicable.

**Informed Consent Statement:** Not applicable.

**Data Availability Statement:** Not applicable.

**Acknowledgments:** The authors gratefully acknowledge the EERA Joint Program Smart Cities.

**Conflicts of Interest:** The authors declare no conflict of interest.

## Nomenclature

| | |
|---|---|
| AI | Artificial Intelligence |
| BMS | Building Management System |
| CNN | Convolutional Neural Network |
| DL | Deep Learning |
| DNN | Deep Neural Network |
| HVAC | Heating, Ventilation and Air Conditioning |
| IAQ | Indoor Air Quality |
| ICT | Information and Communication Technologies |
| IoT | Internet of Things |
| MEL | Miscellaneous Electric Loads |
| R-CNN | Region Based Convolutional Neural Network |
| RH | Relative Humidity |
| Symbols | |
| $C_{CO_2}$ | $CO_2$ concentration (ppm) |
| D | Discomfort parameter |
| E | Energy consumption (kWh) |
| I | Comfort Indicator |
| $I_{set}$ | Set-point value for the indicator |
| $n_{occ}$ | Number of occupants |
| $\dot{r}$ | $CO_2$ production rate per person (mL/s) |
| $t$ | Time (s) |
| $V$ | Volume ($m^3$) |
| $x$ | Variable for the optimum design problem |
| w | Weight for the single objective |

## References

1. Pérez-Lombard, L.; Ortiz, J.; Pout, C. A review on buildings energy consumption information. *Energy Build.* **2008**, *40*, 394–398. [CrossRef]
2. Ash, C.; Jasny, B.R.; Roberts, L.; Stone, R.; Sugden, A.M. Reimagining cities. *Science* **2008**, *319*, 739. [CrossRef] [PubMed]
3. Ciancio, V.; Salata, F.; Falasca, S.; Curci, G.; Golasi, I.; de Wilde, P. Energy demands of buildings in the framework of climate change: An investigation across Europe. *Sustain. Cities Soc.* **2020**, *60*, 102213. [CrossRef]
4. Matthies, E.; Kastner, I.; Klesse, A.; Wagner, H.J. High reduction potentials for energy user behavior in public buildings: How much can psychology-based interventions achieve? *J. Environ. Stud. Sci.* **2011**, *1*, 241–255. [CrossRef]

5. D'Oca, S.; Hong, T.; Langevin, J. The human dimensions of energy use in buildings: A review. *Renew. Sustain. Energy Rev.* **2018**, *81*, 731–742. [CrossRef]

6. Yang, R.; Wang, L. Multi-objective optimization for decision-making of energy and comfort management in building automation and control. *Sustain. Cities Soc.* **2012**, *2*, 1–7. [CrossRef]

7. Malik, M.Z.; Shaikh, P.H.; Khatri, S.A.; Shaikh, M.S.; Baloch, M.H.; Shaikh, F. Analysis of multi-objective optimization: A technical proposal for energy and comfort management in buildings. *Int. Trans. Electr. Energy Syst.* **2020**, *31*, 12736.

8. Djongyang, N.; Tchinda, R.; Njomo, D. Thermal comfort: A review paper. *Renew. Sustain. Energy Rev.* **2010**, *14*, 2626–2640. [CrossRef]

9. Cony, L.R.S.; Abadie, M.; Wargocki, P.; Rode, C. Towards the definition of indicators for assessment of indoor air quality and energy performance in low-energy residential buildings. *Energy Build.* **2017**, *152*, 492–502. [CrossRef]

10. Chui, K.T.; Lytras, M.D.; Visvizi, A. Energy Sustainability in Smart Cities: Artificial Intelligence, Smart Monitoring, and Optimization of Energy Consumption. *Energies* **2018**, *11*, 2869. [CrossRef]

11. Mokhtari, R.; Jahangir, M.H. The effect of occupant distribution on energy consumption and COVID-19 infection in buildings: A case study of university building. *Build. Environ.* **2021**, *190*, 107561. [CrossRef]

12. Jain, N.; Burman, E.; Robertson, C.; Stamp, S.; Shrubsole, C.; Aletta, F.; Barrett, E.; Oberman, T.; Kang, J.; Raynham, P.; et al. Building performance evaluation: Balancing energy and indoor environmental quality in a UK school building. *Build. Serv. Eng. Res. Technol.* **2020**, *41*, 343–360. [CrossRef]

13. Homod, R.Z.; Gaeid, K.S.; Dawoodc, S.M.; Hatamid, A.; Sahari, K.S. Evaluation of energy-saving potential for optimal time response of HVAC control system in smart buildings. *Appl. Energy* **2020**, *271*, 115255. [CrossRef]

14. Li, W.; Wang, S. A multi-agent based distributed approach for optimal control of multi-zone ventilation systems considering indoor air quality and energy use. *Appl. Energy* **2020**, *275*, 115371. [CrossRef]

15. Samadi, M.; Fattahi, J.; Schriemer, H.; Kantarci, M.E. Demand Management for Optimized Energy Usage and Consumer Comfort Using Sequential Optimization. *Sensors* **2021**, *21*, 130. [CrossRef] [PubMed]

16. Shaikh, P.H.; Nor, N.B.M.; Nallagownden, P.; Elamvazuthi, I.; Ibrahim, T. Intelligent multi-objective control and management for smart energy efficient buildings. *Electr. Power Energy Syst.* **2016**, *74*, 403–409. [CrossRef]

17. Shah, A.S.; Nasir, H.; Fayaz, M.; Lajis, A.; Shah, A. A Review on Energy Consumption Optimization Techniques in IoT Based Smart Building Environments. *Information* **2019**, *10*, 108. [CrossRef]

18. Boodi, A.; Beddiar, K.; Benamour, M.; Amirat, Y.; Benbouzid, M. Intelligent Systems for Building Energy and Occupant Comfort Optimization: A State of the Art Review and Recommendations. *Energies* **2018**, *11*, 2604. [CrossRef]

19. Gupta, D.; Mahajan, A.; Gupta, S. Social Distancing and Artificial Intelligence—Understanding the Duality in the Times of COVID-19. In *Intelligent Systems and Methods to Combat Covid-19*; Joshi, A., Dey, N., Santosh, K., Eds.; Springer Briefs in Applied Sciences and Technology: Berlin/Heidelberg, Germany, 2020. [CrossRef]

20. Rezaei, M.; Azarmi, M. DeepSOCIAL: Social Distancing Monitoring and Infection Risk Assessment in COVID-19 Pandemic. *Appl. Sci.* **2020**, *10*, 7514. [CrossRef]

21. Ahmed, I.; Ahmad, M.; Rodrigues, J.J.P.C.; Jeon, G.; Din, S. A deep learning-based social distance monitoring framework for COVID-19. *Sustain. Cities Soc.* **2021**, *65*, 102571. [CrossRef]

22. Saponara, S.; Elhanashi, A.; Gagliardi, A. Implementing a real-time, AI-based, people detection and social distancing measuring system for Covid-19. *J. Real-Time Image Process* **2021**, *21*, 1–11. [CrossRef]

23. Saponara, S.; Elhanashi, A.; Gagliardi, A. Real-time video fire/smoke detection based on CNN in antifire surveillance systems. *J. Real-Time Image Proc.* **2020**. [CrossRef]

24. Aftab, M.; Chen, C.; Chau, C.-K.; Rahwan, T. Automatic HVAC control with real-time occupancy recognition and simulation-guided model predictive control in low-cost embedded system. *Energy Build.* **2017**, *154*, 41–156. [CrossRef]

25. Nagy, Z.; Yong, F.Y.; Schlueter, A. Occupant centered lighting control: A user study on balancing comfort, acceptance, and energy consumption. *Energy Build.* **2016**, *126*, 310–322. [CrossRef]

26. Franco, A. Balancing User Comfort and Energy Efficiency in Public Buildings through Social Interaction by ICT Systems. *Systems* **2020**, *8*, 29. [CrossRef]

27. Rao, S. *Engineering Optimization. Theory and Practice*, 3rd ed.; John Wiley & Sons Inc.: Hoboken, NJ, USA, 1996.

28. Van Hoof, J. Forty years of Fanger's model of thermal comfort: Comfort for all? *Indoor Air* **2008**, *18*, 182–201. [CrossRef] [PubMed]

29. Halawa, E.; van Hoof, J. The adaptive approach to thermal comfort: A critical overview. *Energy Build.* **2012**, *51*, 101–110. [CrossRef]

30. Reiss, J. Energy Retrofitting of School Buildings to Achieve Plus Energy and 3-litre Building Standards. *Energy Procedia* **2014**, *48*, 1503–1511. [CrossRef]

31. Ascione, F.; Bianco, N.; de Masi, R.F.; Mauro, G.M.; Vanoli, G.P. Energy retrofit of educational buildings: Transient energy simulations, model calibration and multi-objective optimization towards nearly zero-energy performance. *Energy Build.* **2017**, *144*, 303–319. [CrossRef]

32. Franco, A.; Leccese, F. Measurement of $CO_2$ concentration for occupancy detection in education buildings with energy efficiency purposes. *J. Build. Eng.* **2020**, *32*, 101714. [CrossRef]

33. Wolf, S.; Calì, D.; Krogstie, J.; Madsen, H. Carbon dioxide-based occupancy estimation using stochastic differential equations. *Appl. Energy* **2019**, *236*, 32–41. [CrossRef]
34. Rahman, H.; Han, H. Occupancy Estimation Based on Indoor CO2 Concentration: Comparison of Neural Network and Bayesian Methods. *Int. J. Air-Cond. Refrig.* **2017**, *25*, 1750021. [CrossRef]
35. American Society of Heating Refrigerating and Air-Conditioning Engineers (ASHRAE), ASHRAE Standard 62.1-2016, Ventilation for acceptable indoor air quality. 2016. Available online: https://www.ashrae.org/File%20Library/Technical%20Resources/Standards%20and%20Guidelines/Standards%20Addenda/62.1-2016/62_1_2016_e_20180126.pdf (accessed on 9 April 2021).
36. European Committee for Standardization (CEN). *EN 13779—Ventilation for Non-Residential Buildings—Performance for Ventilation and Room Conditioning Systems*. 2015. Available online: https://www.rehva.eu/rehva-journal/chapter/ventilation-for-non-residential-buildings-performance-requirements-for-ventilation-and-room-conditioning-systems (accessed on 9 April 2021).
37. Becchio, C.; Corgnati, S.P.; Delmastro, C.; Fabi, V.; Lombardi, P. The role of nearly-zero energy buildings in the transition towards Post-Carbon Cities. *Sustain. Cities Soc.* **2016**, *27*, 324–337. [CrossRef]

*Article*

# PEDRERA. Positive Energy District Renovation Model for Large Scale Actions

Paolo Civiero [1,*], Jordi Pascual [1], Joaquim Arcas Abella [2], Ander Bilbao Figuero [2] and Jaume Salom [1]

[1] Thermal Energy and Building Performance Group, IREC—Catalonia Institute for Energy Research, Sant Adrià del Besos, 08930 Barcelona, Spain; jpascual@irec.cat (J.P.); jsalom@irec.cat (J.S.)

[2] CÍCLICA ARQUITECTURA SCCL, Sant Cugat del Vallès, 08173 Barcelona, Spain; joaquim.arcas@ciclica.eu (J.A.A.); ander.bilbao@ciclica.eu (A.B.F.)

* Correspondence: pciviero@irec.cat

**Abstract:** In this paper, we provide a view of the ongoing PEDRERA project, whose main scope is to design a district simulation model able to set and analyze a reliable prediction of potential business scenarios on large scale retrofitting actions, and to evaluate the overall co-benefits resulting from the renovation process of a cluster of buildings. According to this purpose and to a Positive Energy Districts (PEDs) approach, the model combines systemized data—at both building and district scale—from multiple sources and domains. A sensitive analysis of 200 scenarios provided a quick perception on how results will change once inputs are defined, and how attended results will answer to stakeholders' requirements. In order to enable a clever input analysis and to appraise wide-ranging ranks of Key Performance Indicators (KPIs) suited to each stakeholder and design phase targets, the model is currently under the implementation in the urbanZEB tool's web platform.

**Keywords:** Positive Energy District; smart districts; building performance simulation; sustainable large-scale renovation model; Driving Urban Transition; Renovation Wave

**Citation:** Civiero, P.; Pascual, J.; Arcas Abella, J.; Bilbao Figuero, A.; Salom, J. PEDRERA. Positive Energy District Renovation Model for Large Scale Actions. *Energies* **2021**, *14*, 2833. https://doi.org/10.3390/en14102833

Academic Editor: Alfonso Capozzoli

Received: 26 April 2021
Accepted: 11 May 2021
Published: 14 May 2021

**Publisher's Note:** MDPI stays neutral with regard to jurisdictional claims in published maps and institutional affiliations.

## 1. Introduction

The European Commission has set ambitious targets to make Europe the first carbon-neutral continent by 2050. As the building sector is one of the largest energy consumers, the European Union (EU) is now stepping up efforts towards citywide transformation to enable transitions towards a climate neutral economy. A refurbished and improved building stock in the EU will help to pave the way for a decarbonized and clean energy system as well as for the development of neutral cities [1]. The improvement of the intervention rate up to 3% per year, which means the need to promote renovation actions, will raise the overall quality of the building stock, especially regarding the energy neutrality, high efficiency and health [2,3]. Indeed, large-scale renovation means also to regenerate and revitalize the social and economic structures locally, and to build trust in the business opportunities and benefits for each actor involved in the process [4]. To pursue these ambitious energy and climate benefits and the economic growth, the Positive Energy Districts (PEDs) approach will be a driving force, bringing the European Green Deal (GD) closer to citizens in an attractive, innovative and human-centered way [3]. The pioneering concept of PEDs, which builds on the paradigm of smart cities, will be incrementally introduced in the energy planning of many cities and communities in the coming years [5]. The Positive Energy District is directly contributing to the Renovation Wave through the strengthening of national innovation policies by coordinating, pooling and increasing of Research and Innovation funding for developing 100 Positive Energy Districts in Europe by 2025. Furthermore, PEDs will help to fulfil the goals set out by international policy frameworks such as the Urban Agenda for the EU, the COP21 Paris Agreement, the Habitat III New Urban Agenda, and the UN's Sustainable Development Goals (notably SDG 11), and to boost the large-scale regeneration of the built environment.

Aligned with this perspective, the paper describes the ongoing research project "PE-DRERA. Positive Energy District renovation model" whose main goal is to handle several key challenges and sectoral priorities of this urban transitions, by unlocking the potential of business models geared towards large scale refurbishment plans [6]. A model able to provide a reliable prediction of potential scenarios—and their benefits in terms of energy efficiency, well-being and economic topics, among others—in order to plan and execute investment. Indeed, the purpose of the hereunder described PEDRERA project is to create a multidimensional urban building energy modeling (UBEM) [7] tool able to assess and promote the large-scale renovation in the urban areas. According to this objective, the model supports the simulation of different renovation scenarios moving from a set of information that firstly is automatically gathered and/or extracted. The definition of the main input and the calculation of Key Performance Indicators (KPIs) for large-scale renovation strategies are proposed according to a PEDs vision including different stakeholders' perspectives and with special attention to the most vulnerable groups, as well to make the transition perspective, feasible. Once collected, inputs are used to run the PEDRERA model algorithms and suited on each stakeholder involved in the renovation process. In this way, it will promote an action of urban regeneration focused on clear long-term environmental, social and economic objectives [8].

The present paper, oriented to the problem definition and formulation, ready for the implementation, demonstrates how the research addresses several challenges which represent some of the main barriers for this transition process, especially:

- How to achieve and integrate heterogeneous data from different domain in order to get a comprehensive understanding of district scale renovation complexity.
- How the information is systematized and addressed to specific KPIs and to stakeholders' targets along the renovation process phases.

Although the potential of available data, some critical barriers could hinder their effectiveness and implementation of the PEDs in the urban environment, and are represented by the ability to aggregate data from different sources and to exploit this information according to specific target groups' "scope". Indeed, the challenge is not only how to gather fair data (reliable, verified and continuous over time), but also how to integrate them in order to formulate predictive and feasible business models adherent to the purpose. One of the key innovation aspects of the PEDRERA model means the way to solve the gaps and the integration from different databases and dispersed information.

According to these addressed challenges, it is well known that the availability of widely monitored and shared data is certainly one of the key aspects of smart and digital cities [9]. Aimed to facilitate collecting, sharing and integrating data about the environment, several initiatives—like the OECD's Open Government Data project and the INSPIRE directive (currently under revision and that will be implemented by 2021)—are meant to create a European Union spatial data infrastructure for the purposes of EU environmental policies [10,11]. In compliance with these initiatives, national land registers—such as the Spanish cadaster—are already sharing their spatial data, including parcels and building characteristics (gross surfaces, number of dwellings and of floors), among others.

On the other hand, in order to address information to specific KPIs and thus explore different scenarios and potentialities allowed, the PEDRERA model adopts the innovative methodology of UBEM developed in the urbanZEB tool provided by the CICLICA, partner of the research project [12,13]. UBEMs have emerged in recent years as efficient hybrid of top-down statistical and bottom-up engineering approaches [14–17] and they are expected to become the main planning tool for energy utilities, municipalities, urban planners and other professionals [18–21]. Both the UBEMs and the virtual city maps are considered as the new generation of tools that based on the digital twins concept, allow the analysis and monitoring of large urban areas and built stock [22–25]. The UBEM allows the multilevel integration of several sources of information, the energy simulation of building and characterization of buildings, and as a result, the generation of new essential knowledge and new scenarios for urban regeneration [26–28].

Two main UBEM approaches can be identified from the literature: physical modelling and data-driven modelling that provide automated generation of building energy models through abstraction of building stock by different "building archetypes", i.e., sample or virtual buildings that characterize subsets of buildings of the same kind [29,30]. With regards to the other existent tools, the joint venture PEDRERA and urbanZEB tool leads to a very dynamic, flexible and definitively accurate data-driven UBEM tool for large-scale prediction. Furthermore, PEDRERA is benchmarking innovative and adaptable refurbishment packages on buildings and supporting the design of successful and effective business models for their large-scale deployment and replication. Based on a quantitative aggregation of data, obtained from different sources (e.g., cadaster, public energy performance certificates, statistical socio-economic local conditions, etc.) the adopted software engine algorithms are already able to both support the calculation of different packages of intervention and provide the simulation of potential scenarios, while dealing with different energy goals (energy efficiency and production), as well costs savings and environmental/welfare co-benefits.

Together these models also enable the analysis of the energy supply and demand in a region; make it possible to develop scenarios; determine a preferred mix of technologies, given certain constraints; simulate behavior of energy producers and consumers in response to prices and other signals, etc.

Nevertheless, based on the lesson learned from similar existent models and software, PEDRERA project aims to build a user-friendly, easy replicable and updatable model that could be useful for energy system transformation and supporting PEDs implementation. For this reason, the number of output (KPIs) handled by the model is reduced to the most relevant information (Table 1). Furthermore, the KPIs are strictly based on each type of actor of the process, thus when specific aspects (e.g., socioeconomic ones) mean the main issue to be considered in the process scenario, prioritization ranking will help to select both the crucial inputs and the drivers to be adopted.

The urbanZEB platform, where PEDRERA model will be implemented, is currently in use and has been adopted to define and support local and national plans and strategies. The results of urbanZEB are accessible through an interactive online platform that allows to consult both the single building level and multiple buildings through mapping, based on graphical data, which include functionalities for comparing scenarios and spatial geographic filtering. The visualization of urban information is in three possible output formats: 3d cartography, database and tables and graphs, while the consultation of urban information is according to three spatial units of analysis: building, census section or neighborhood and municipality. The most relevant experience of implementing urbanZEB so far was the long-term renovation strategy (LTRS) in the Basque Country's building stock, in 2019. The project challenged an innovative action plan which, for the first time on a regional scale, was based on the building-by-building diagnosis of 1.1 million dwellings, providing a significant advancement in the methodology so far employed for large-scale renovation strategies. At metropolitan scale, the urbanZEB tool was implemented in the Barcelona Metropolitan Area, in order to prioritize the intervention of energy renovation in an area of special vulnerability, formed by more than 200,000 dwellings. The urbanZEB tool's calculation engine was also used in the Spanish LTRS 2020 for the energy simulation of the archetypes identified, with the aim of obtaining the energy reduction after the intervention in the national stock.

Once implemented, the PEDRERA tool will facilitate the engagement of multiple stakeholders involved in the building renovation programs to make effective and well-informed decisions from a cluster of georeferenced buildings. Indeed, the urbanZEB tool is able to simulate each of the sub-models implemented, and where it is possible to choose among many existing simulation engines or tailor-made models, which fit the characteristics of an application case. Hence, the integration and systematization of the information from multiple domains (e.g., building, energy, economy, financing) is the first step to assess and to manage accurately both the complexity of financing renewal processes at district scale and the interests of each stakeholder. The bottom-up approach of the urbanZEB tool, dealing with the numerical representation of interconnections between the buildings and the surrounding environment, can assess the needs of several stakeholders—including the final users' requirement—with a high resolution of outputs. In that case of the PEDRERA model, the accuracy of the simulation engine has been tested and validated to forecast potential scenarios with results that adhere to the market values.

## 2. Methods

The methodology deployed in this analysis is summarized in Figure 1. The tool framework is drawn according to a step by step interactive approach where a different deepening of input belongs to each phase of the renovation process: (1) data aggregation, (2) leading phase, (3) demand aggregation, (4–5) concept and technical design, (6) construction, (7) use. As described above, the PEDRERA data-driven model is based firstly on the aggregation of the information system that means to: (1) collect and gather the available data from multiple domains; (2) integrate the available data to create data-driven models and scenarios that enable to gain a better understanding of the complex reality—extended to no-directly related building stock information.

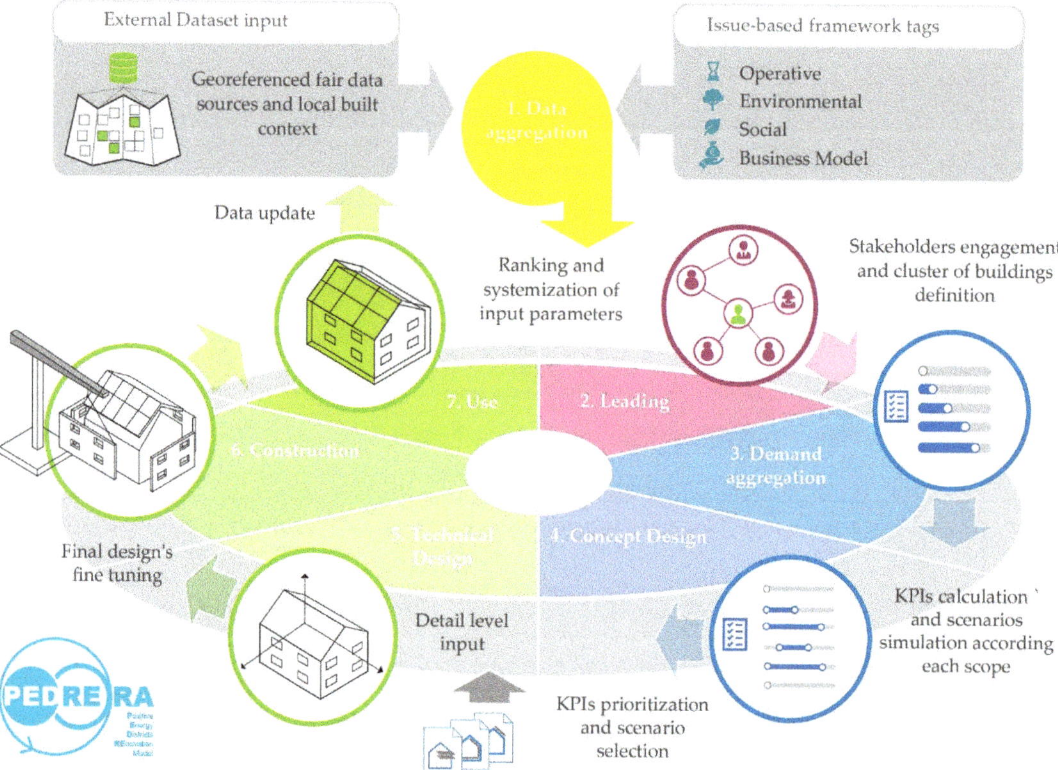


**Figure 1.** Synthesis of the renovation process flowchart and the interactive approach with the PEDRERA model. Source: PEDRERA.

According to the interactive approach scheme presented in Figure 1, the data aggregation from multiple domains and sources (**Step 1—data aggregation**) is arranged in a four main domains framework that will be used to assess the renovation program: business model, environmental, operative and social issues (*tags*).

Indicators to evaluate the status renovation potential of a particular urban area are supposed to capture and process fair data from multiple sources: energy performance, census, cadaster, building physical characteristics, etc. Each input means an indicator that will be introduced in the algorithms for the calculation of the different KPIs according to the scopes:

- Financial appraisal.
- Renovation strategy.
- Energy community.
- Welfare and security.
- Marketing.

Moreover, socio-economic indicators such as low income and aging population are also observed in order to cluster most intensely certain urban environments as representative of social cohesion as well as useful to design the business models and to evaluate their impact on each scenario. Since they are significantly representative of household vulnerability, they are likewise considered as input of the risk poverty for both the sustainability of the business model, and the renovation strategy to adopt. That information brings together the first collection of input that will be used as basis of knowledge and for the design of the renovation scenarios. For this scope, the different types of sources are defined and listed. Otherwise, inputs are imported, calculated and/or simulated from statistical information for single customized projects.

The aggregated data are systemized according to the sub-categories and scopes of the PEDRERA framework, as described later. All the information is georeferenced to the single buildings according to the above mentioned four "tags" (Figure 2) that make quick and clever the selection and the assessment of potential buildings to be renovated, as well as taking into account the engagement of the stakeholders (agent/user) involved in the process (**Step 2—leading**) (Figure 3).

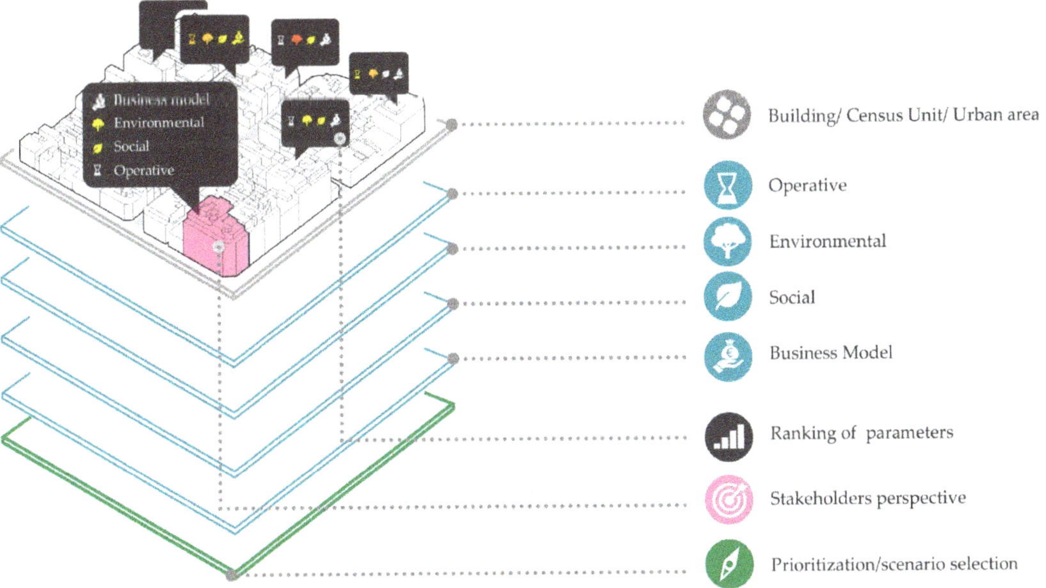


**Figure 2.** Multidimensional model indicators (layers) displayed in the platform and supporting residential building retrofitting programs. Source: authors.

1. Data aggregation
2. Leading

(a)

(b)

**Figure 3.** Extract of the PEDRERA model approach expected visualization in the urbanZEB tool: (**a**) Data aggregation at urban level; (**b**) leading phase and stakeholder engagement for the cluster of building selection. Source: authors.

Input data are supposed to be automatically gathered from a large number of external data sources parameters (e.g., technical and architectural aspects, weather, demographic/aging condition, income/energy poverty indicators or socio-economic rating, energy demand and consumption data, etc.). Multiple sources of data are available for this scope (open data) and can be freely used, shared and built-on by anyone, anywhere, for any purpose, although each one is managed by a different organization, and collected for a specific purpose (e.g., improving building energy efficiency, evaluating the status of the building stock, etc.). In addition to gathering and integrating the aggregated data, the PEDRERA model algorithm exploits the potential relationships between data to design each renovation measure and to evaluate their impact on all the above listed scopes (**Step 3—demand aggregation**) (Figure 4).

3. Demand aggregation

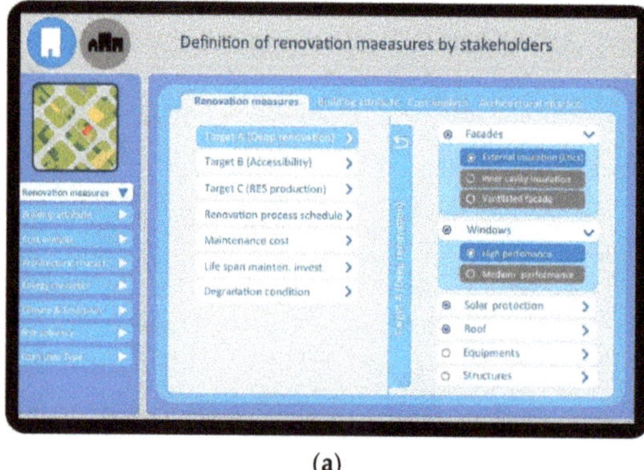

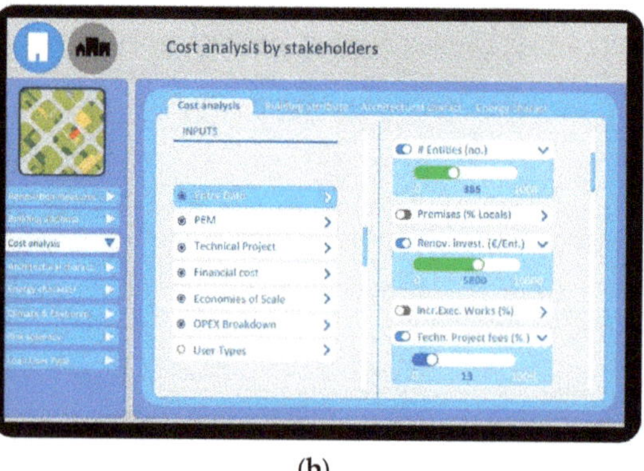

(a)

(b)

**Figure 4.** Extract of the PEDRERA model approach, expected visualization in the urbanZEB tool. Demand aggregation menu with: (**a**) the renovation measures input; (**b**) the integration of aggregated and gathered data for the cost analysis. Source: authors.

Once inputs are collected, systemized and filtered, the model provides the calculation of composite performance output (KPIs), as well as the information on compliance with the targets established by the stakeholders. Both the value and the prioritization rank of each input (attributes and variables to be included in the algorithms) and output KPIs will modify the renovation strategy to be adopted and, consequently, the scenarios results (**Step 4—concept design**). The forecasted scenarios displayed do reflect almost a range of the potential menus of intervention and, thanks to their friendly visualization, will help the exploitation of the results obtained for the next design phases (**Steps 5, 6 and 7 of the renovation process**), integrating the stakeholders' requirement and expectations (Figure 5). After the conclusion of the renovation process, the same platform will also enable to update data stored in the repository for future analysis and intervention.

**4. Concept Design**

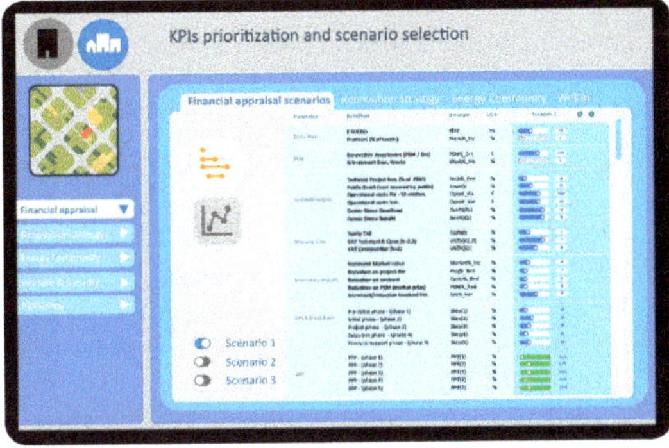

(a)

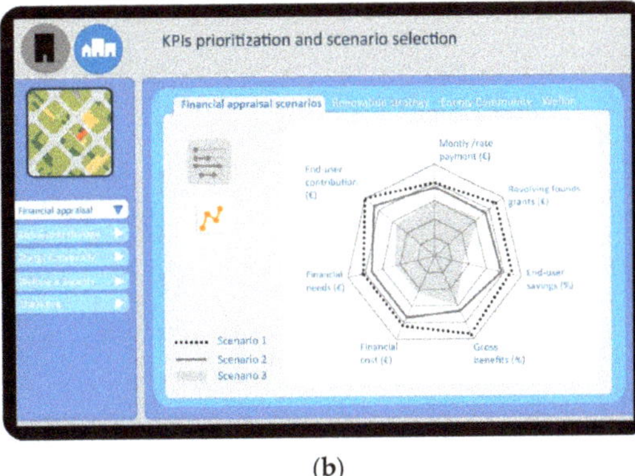

(b)

**Figure 5.** Extract of the PEDRERA model approach, expected visualization in the urbanZEB tool: (**a**) concept design based on financial appraisal and KPIs prioritization; (**b**) scenario visualization for scenario selection according to priorities from stakeholders. Source: authors.

*PEDRERA Model Input*

Table 1 provides a summary of the inputs required by the PEDRERA model to properly run simulations. Most of the input data for the algorithms are already collected from public sources and from case study sources and integrated in the urbanZEB tool (e.g., cadaster, energy performance certificates and census, among other) by Semantic Web processing.

Each input is systemized in the model framework in order to predicts the suitable output (KPIs) for given inputs. The input's features are namely: type of **issue** and type of **parameter**, **indicator**, **unit**, **scope** (the goal where input is adopted), **data source** (imported from database, calculated, simulated), **scale** (building, district, urban area/census unit) and **source dataset**.

Therefore, the PEDRERA model will run data through a special wizard able to filter input data which are already displayed and shared on the urbanZEB web platform. Through Geographical Information System (GIS), Big Data and Extract, Transform and Load (ETL) techniques these sources are integrated in a Data Warehouse that allows the dynamic crossing of the different levels of information. The connectors will be implemented with API calls, data dump digestion or in web scrapping techniques based on the characteristics of the sources. In this way, the processes are configured to cover specific case studies on extended urban areas. The repository is implemented as a PostgreSQL/PostGIS database in order to facilitate displaying the data in map visualizations. Data sources that are not published in a structured format such as CSV, RDF, API REST and XLS, among others, are processed using web scrapping techniques. The database file stores all the information resulting from the characterization phase of the project (with all the parameters and indicators) and the prioritization indices. In addition, to facilitate the reading and visualization of the data, an X-ray file (ArcMap) holds the layers of information established in the project and their set of symbols.

**Table 1.** List of main input implemented and managed in the PEDRERA model. Source: authors.

| Issue | Parameter | Indicators | Unit | Scope | | | | | Data Source | | | Scale | | | Source Dataset |
| --- | --- | --- | --- | --- | --- | --- | --- | --- | --- | --- | --- | --- | --- | --- | --- |
| | | | | Financial Appraisal | Renovation Strategy | Energy Community | Welfare & Security | Marketing | Imported from Data base | Calculated | Simulated | Building | District | Urban area / Census unit | |
| Business model | Building attributes | Property (public/private ownership) | % | ● | ● | ● | | | ● | | | ● | | | National cadaster |
| | Building attributes | Property ratio (Owners/tenants ) | % | ● | | ● | | | ● | ● | ● | ● | | | Comdomium/Insurance company |
| | Building attributes | Entities/Dwellings | no. | ● | | | | | | ● | ● | ● | ● | | National cadaster |
| | Economies of Scale | Reduction on contract (tender phase) | % | ● | ● | | | | | ● | ● | ● | ● | | PEDRERA model +ITEC |
| | Economies of Scale | Reduction on PEM (market prize) | % | ● | ● | | | | | ● | ● | ● | ● | | PEDRERA model +ITEC |
| | Economies of Scale | Increment/Reduction on involved entities | % | ● | ● | | | | | ● | ● | ● | ● | | PEDRERA model |
| | Economies of Scale | Reduction on project fees | % | ● | ● | | | | | ● | ● | ● | ● | | PEDRERA model |
| | Economies of Scale | Increment Market value | % | ● | ● | | | | ● | | ● | ● | ● | | Incasol/AMB /PMRH |
| | Cost analysis | VAT (Technical + Operational) | % | ● | | | | | ● | | | ● | ● | | PEDRERA model |
| | Cost analysis | Financing amount | € | ● | | | | | | | ● | ● | ● | | PEDRERA model |
| | Cost analysis | Operational Cost Breakdown | % | ● | ● | | | | | | ● | ● | ● | | PEDRERA model |
| | Cost analysis | PPP ratio according renovation process | % | ● | | | | | | | ● | ● | ● | | PEDRERA model |
| | Cost analysis | Technical Project fees | % | ● | ● | | | | ● | ● | ● | ● | | | PEDRERA model |
| | Cost analysis | Operational costs (Fix. + Var.) | € | ● | ● | | | | | ● | ● | ● | ● | | PEDRERA model |
| | Cost analysis | OverHeads | % | ● | ● | | | | | ● | ● | ● | ● | | PEDRERA model |
| | Cost analysis | Benefits | % | ● | ● | | | | | ● | ● | ● | ● | | PEDRERA model |
| | Cost analysis | Public Grant (costs covered by public) | % | ● | ● | | | | | ● | ● | ● | ● | | ERDF/National EE Fund/PAREER II/ IDAE |
| | Cost analysis | Loans years | no. | ● | | | | | | | ● | ● | ● | | PEDRERA model |
| | Cost analysis | Yearly TAE | % | ● | | | | | | | ● | ● | ● | | PEDRERA model |
| | Cost analysis | Renovation investment (PEM/Ent.) | € | ● | ● | | | | | | ● | ● | ● | | PEDRERA model |
| | Cost analysis | Maintenance cost (costr. elements) | % | ● | ● | | | | | ● | ● | ● | | | GBCe + BEDEC/CYPE |
| | Cost analysis | Maintenance cost (equipments) | % | ● | ● | | | | | ● | ● | ● | | | GBCe + BEDEC/CYPE |
| | Cost analysis | Life span maintenance investment | years | ● | ● | | | | | | ● | ● | | | PEDRERA model |
| | User Types | UT A (A1, A2, A3, … ) | % | ● | | | | | | ● | ● | ● | | | PEDRERA model |
| | User Types | UT B (vulnerable persons) | % | ● | | | | | | ● | ● | ● | | | PEDRERA model |
| | User Types | UT C (Defaulters) | % | ● | | | | | | ● | ● | ● | | | PEDRERA model |

**Table 1.** *Cont.*

| Issue | Parameter | Indicators | Unit | Scope | | | | | Data Source | | | Scale | | | Source Dataset |
| | | | | Financial Appraisal | Renovation Strategy | Energy Community | Welfare & Security | Marketing | Imported from Data base | Calculated | Simulated | Building | District | Urban area / Census unit | |
| --- | --- | --- | --- | --- | --- | --- | --- | --- | --- | --- | --- | --- | --- | --- | --- |
| Operative | Architectural character. | Floors | no. | ● | ● | ● | | | ● | | | ● | | | National cadaster |
| | Architectural character. | Roof extension | Sqm | ● | ● | ● | | | ● | ● | ● | ● | ● | ● | National cadaster + AMB/DIBA |
| | Architectural character. | Envelope extension | Sqm | ● | ● | ● | | | ● | ● | ● | ● | | | UrbanZEB tool |
| | Architectural character. | Facades ratio | % | | ● | ● | | | | ● | ● | ● | | | UrbanZEB tool |
| | Architectural character. | Windows ratio | % | | ● | ● | | | | ● | ● | ● | | | UrbanZEB tool |
| | Architectural character. | Building cluster type (Age) | - | | ● | ● | | | ● | | | ● | | | UrbanZEB tool |
| | Architectural character. | Building cluster type (Use) | - | | ● | ● | | | ● | | | ● | | | UrbanZEB tool |
| | Building attribute | Residential/tertiary | % | ● | | ● | | | ● | | | ● | ● | | National cadaster/ Comdomium/Insur. company |
| | Building attribute | Accessibility | - | | ● | | ● | | ● | | | ● | | | AHC/comdomium/ Insurance company |
| | Building attribute | Degradation Condition | Range | ● | ● | | ● | ● | ● | | | ● | ● | | AHC/comdomium/ Insurance company |
| | Energy character. | Energy Demand | - | | | ● | ● | | ● | | | ● | ● | | AMB/DIBA |
| | Energy character. | Final energy consumption | - | | | ● | ● | ● | ● | | | ● | ● | | AMB/DIBA |
| | Renovation measures | Renovation targets A (Env. + Equip. + Struct.) | - | | ● | ● | ● | ● | | ● | ● | ● | ● | | PEDRERA model/UrbanZEB tool |
| | Renovation measures | Renovation targets B (Accessibility) | - | | ● | ● | ● | | | ● | ● | ● | ● | | PEDRERA model/UrbanZEB tool |
| | Renovation measures | Renovation targets C (RES production) | - | | ● | ● | ● | ● | | ● | ● | ● | ● | | PEDRERA model/UrbanZEB tool |
| | Renovation measures | Renovation process (TimeSeries) | Months | ● | ● | | | | | ● | ● | ● | ● | | PEDRERA model |
| Environmental | Climate & Environment | Solar irradiance (horizontal surface) | W/Sqm | ● | ● | | | | ● | ● | | ● | ● | | District modelization/ Solar atlas |
| | Climate & Environment | Solar irradiance (vertical surfaces) | W/Sqm | ● | ● | | | | ● | ● | | ● | ● | | District modelization/ Solar atlas |
| | Energy character. | Electric consumption | kWh/y | ● | ● | | | | ● | | | ● | ● | | Energy Traders |
| | Energy character. | Installable PV kWp factor | % | ● | ● | ● | | ● | ● | ● | ● | ● | ● | | PEDRERA model/UrbanZEB tool |
| | Energy character. | Ratio PV production | Sqm/kW | ● | ● | ● | | ● | ● | ● | ● | ● | ● | | PEDRERA model/UrbanZEB tool |
| | Energy character. | Required kW | kW | ● | ● | ● | | ● | ● | ● | ● | ● | ● | | PEDRERA model/UrbanZEB tool |
| | Energy character. | Boundary of the EC (Radius < 500 m) | m | ● | | ● | | | | ● | ● | ● | ● | | UrbanZEB tool |

**Table 1.** *Cont.*

| Issue | Parameter | Indicators | Unit | Financial Appraisal | Renovation Strategy | Energy Community | Welfare & Security | Marketing | Imported from Data base | Calculated | Simulated | Building | District | Urban area / Census unit | Source Dataset |
|---|---|---|---|---|---|---|---|---|---|---|---|---|---|---|---|
| | Energy character. | Energy Vector | - | | • | • | | | • | | | • | • | | Energy Traders |
| | Energy character. | EPC | - | | • | • | | | • | | • | • | | | ICAEN |
| | Energy character. | Primary energy consumption | - | | • | • | | • | • | | • | • | | | AMB/DIBA |
| Social | User Types | End-user age range | % | • | | • | • | | • | | • | | | • | INE Census |
| | User Types | Households age composition | % | • | | • | | | • | | • | | | • | INE Census |
| | User Types | Gross disposable households incomes | % | • | • | • | • | | • | | | | | • | INE Census |
| | User Types | People at risk of energy poverty | % | • | • | | • | | • | | • | | | • | INE Census + UrbanZEB tool |
| | User Types | Unemployment | % | • | | • | • | | • | | | | | • | INE Census |

The inputs are organized according to both tags and specific parameters, and refer to special feature types adopted in the model engine to calculate the KPIs: building attributes, cost analysis, user types, architectural characterization, energy characterization, renovation measures, economies of scale, climate and environment (Figure 6).

As described above, one of the main information required for the definition of potential scenarios is represented by the architectural characterization feature type of the residential stock. For this purpose and with the aim of simplifying the assessment of building performance in large-scale retrofitting programs, different building archetypes are used (Figure 7 and Table 2).

The description of archetypes, which comes from the methodology used in the national Grupo de Trabajo sobre Rehabilitación (GTR) 2011 report and the LTRS 2020—set out in the EPBD 2010/31/EU—allows the understanding of the residential stock based on the segmentation in groups of buildings that present similar conditions and therefore require similar intervention actions [31]. This characterization is focused on setting the parameters that will have the greatest impact on its energy performance, as well as its related weakness, in order to design the best intervention strategy and assess the economic investment required. This phase is mainly based on processing extraction of cadastral data and is available throughout the urban classifications and by generating more than 300 cross variables related to: location, use and areas, type of residential property, year of construction, number of floors, number of dwellings.

In addition, the collected data include the performance of each building element and envelope—facades, internal wall, roofs, floors and ground floors—as well as the incidence form, the windows performance, rate, size and type. In this way it is possible to carry out the geometric modelling of each building, and to define the features and surfaces of each level on façades and roof, detaching patios and interferences with neighboring or surrounding buildings (e.g., shadows cast) that could be adopted for both the energy demand and the renewable energy sources (RES) production KPIs.

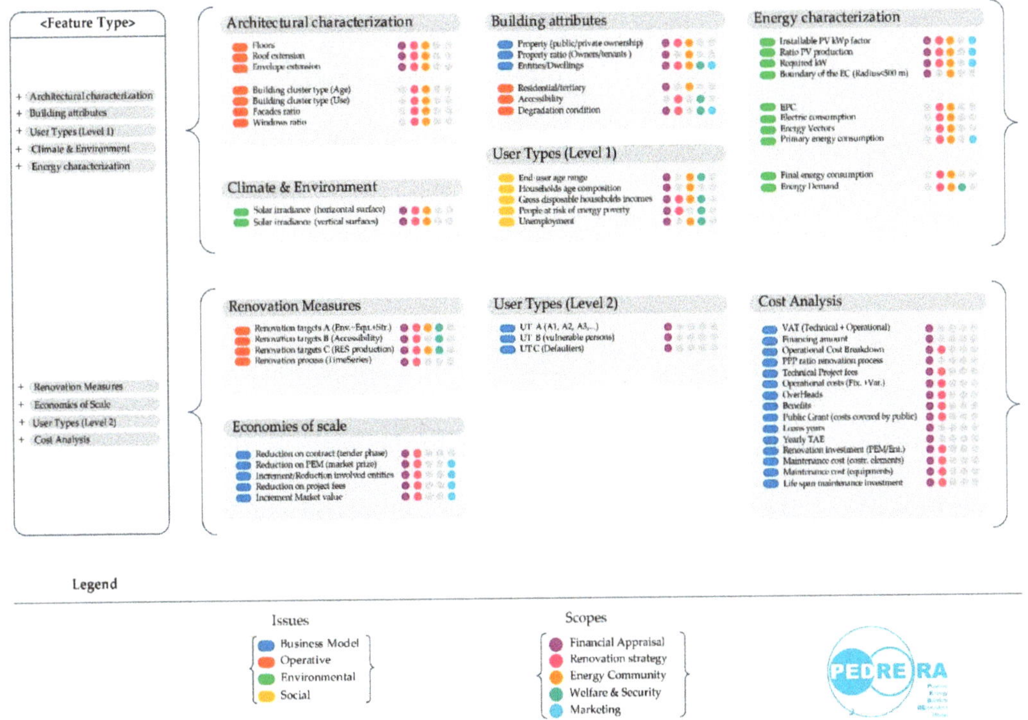



**Figure 6.** Main input implemented and managed in the PEDRERA model framework. The inputs are organized in compliance with the four issues, and each of them is adopted in the algorithms for the evaluation of KPIs according to the scopes. Source: PEDRERA.

**Figure 7.** Aerial view showing the filters applied on residential buildings, and presenting the 12 archetypes based on year of construction and typology. Source: urbanZEB web platform.

**Table 2.** List of main building archetypes implemented and managed in the PEDRERA model. Source: urbanZEB web platform.

| | Unifamiliar | Plurifamiliar |
|---|---|---|
| before 1900 | A | G |
| 1901–1940 | B | H |
| 1941–1960 | C | I |
| 1961–1980 | D | J |
| 1981–2007 | E | K |
| 2008–2020 | F | L |

An overall overview of the construction systems is also made for each building according to the classification into clusters, thus displaying the current building state and enabling the design of post-intervention building scenarios resulting by the application of each intervention menu provided by the model (i.e., input from renovation measures A, B and C as shown previously) (Figure 4, Table 1) within the feature type **renovation measures**. Simultaneously with the architectural characterization parameters, the **energy characterization** inputs are focused on defining the current energy level and the potential from reducing the demand and the energy consumption through the intervention for single buildings. As explained above, for the determination of this set of input, the reference values of the national GTR Reports are used as defined by the Technical Building Code (CTE) and the Institute for Diversification and Energy Saving (IDAE) [2].

For the calculation of energy costs and CO2 emissions of materials and intervention belonging to a typological cluster, the Catalan database of construction elements (BEDEC) from the Institut de Tecnologia de la Construcció de Catalunya (ITeC) is available and adopted. Furthermore, according to a PEDs perspective, and the positive energy balance referred to a cluster of buildings (Building Portfolio) at the neighborhood scale, the calculation of the energy performance (EP) follows the overarching framework of the EN ISO 52000-1:2017. The new EN ISO 52000 family of standards to assess the EP of buildings offers a great flexibility in the calculation according to different choice of assessment methods defined [32]. Otherwise, a stochastic model supports a wide volume of simulations from the very beginning of the renovation process by using statistical information [33].

A territorial **socio-economic index** (IST) from Catalonia region provided by the IDESCAT [34] is adopted. This information, based on the Spanish National Institute of Statistics (INE) census, means a synthetic index by small areas (census units) that aggregates in a single value several socio-economic characteristics of the population.

That information is very relevant not only for the evaluation of the economic effort allowable by the users or the revolving funds required to be covered by the Public Sector, but also to predict the impact of the user' profile on electric consumes or load while designing PEDs. The index concentrates information on the employment situation, educational level, immigration and income of all people living in each territorial unit, based on 6 sectoral indicators. The IST is a relative index, with no units of measurement. A reference value for Catalonia region is established equal to 100, thus each unit is valuable in comparison with this average value. Values per decile are also referred: the first decile refers to the areas with the lowest socio-economic level and the tenth decile to that ones with the highest level.

The PEDRERA model provides a list of **renovation measures**—integrated in the renovation strategy menu—that, based on operative, environmental, economic and social input gathered from the selection of buildings, will enable users to evaluate the estimative metric computation, but also to simulate the energy improvement on the building according to established menu of intervention already available within the model, as well as the other benefits on welfare and security. The renovation measures included in the PEDRERA model are functional in the large-scale refurbishment projects for every cluster of residential buildings selected, as previously described (Figure 1). The menu of intervention refers to 3 main targets of intervention, with a specific type of solution that can be implemented or replaced within the residential buildings:

- Renovation target A (Deep renovation): facades, windows, solar protection, roof, equipment (boilers, heat pumps, lighting, … ), structures reinforcements.
- Renovation target B (Accessibility): ramps, lifts and elevators.
- Renovation target C (RES individual/community): solar and PV panel.

With regards to the **cost analysis**, the inputs are mainly managed within the model instead to be gathered from an external data source. Inputs are finalized to provide specific information on potential **investment KPIs** and several co-benefits to be achieved through the renewal process, and that are profitable and/or feasible for the actors involved in the process. Hence, the building renovation assessment does address and disclose the co-benefits from the renovation measures adopted and the economies of scale, the RES production, the appreciation in housing value, the increase of building lifespan, the impact on the environment and the improved health benefits for both the households and the healthcare system [28]. At the same time, the data and related benefits motivate and empower all stakeholders and target groups to do action. In this way, the city planners can focus on the most beneficial areas in both energy and quality renovation efforts, while the investors can get better access on the information of building stock, which eases making the financial decisions related to large-scale renovation projects benefits and risks. Then, data increase building owner's interest on the performance of their buildings when they can compare their consumption and renovation need, and they can see the potential derived from the renovation and the related businesses of energy improvements (e.g., the estimation if the roof renovation can be combined with PV panel installation and the benefits to establish an energy community) as shown in Table 3.

## 3. Results

Preliminary results obtained by the ongoing PEDRERA project refer to the definition of the conceptual framework of the model and in regards to: (a) data sources aggregation according to the four domains described above; (b) input required for the KPIs calculation (Table 1) that can be assumed to assess different "scopes". Aligned with this vision, the model is powered by the integration of the processed input for the calculation of the most relevant KPIs (outputs) algorithms according to each process phase and stakeholder's profile (Table 3).

*PEDRERA Model Ouput*

Table 3 summarizes the outputs provided by running a typical simulation with urbanZEB tool powered by the PEDRERA model. It must be noted that the outputs are assigned to five main targets related with the **scopes**: (i) financial appraisal, (ii) renovation strategy, (iii) energy community, (iv) welfare and security, (v) marketing.

In order to complete the panel of KPIs requested from large actions on building stock, the project focuses also on the perspectives of selected target groups that can be considered as key actors for urban development. The main considered stakeholders that would manage scenarios from the data-driven model are mainly:

- **End users:** the owners and tenants, intervening as individuals or as members of communities.
- **Public authorities:** the local and/or regional administrations, responsible of the planning and building regulations and partners in public–private partnerships.
- **Financial institutions:** banks, investors and ESCO companies, which provide private funds for building renovation programs.
- **Property developers:** real estate agencies, building administrators, home insurance companies or other stakeholders (e.g., one-stop shop) attracted from an investment or operational point of view.

**Table 3.** List of output (KPIs) of the PEDRERA model according to the "scopes" and stakeholder engagement for each phase of the process. Source: authors.

| Outputs (KPIs) | Acronym | Units | Phase | | | | Stakeholder | | | | | Scope | | | |
|---|---|---|---|---|---|---|---|---|---|---|---|---|---|---|---|
| | | | Leading phase | Demand aggregation | Concept & Techn. design | Construction & Use | End-users | Public Authorities | Financial institutions | Property developers | Financial Appraisal | Renovation Strategy | Energy Community | Welfare & Security | Marketing |
| End-users contribution | UserCont | € | ● | ● | | | ● | | | | ● | ● | | | |
| Monthly/rate payments | UserPaym | € | | ● | | ● | ● | | | | ● | | | | |
| End-users savings | UserSave | % | | ● | | | ● | | | | ● | | | | |
| Revolving fund grants | RevGrant | € | ● | ● | | | | ● | | | ● | | | ● | |
| Operational Costs | OpCost | € | ● | ● | | | | ● | ● | ● | ● | | | | |
| Early Before Taxes | EBT | € | | ● | ● | | | | | ● | ● | | | | |
| Gross. Benefits | GrossBen | % | ● | ● | ● | | | | | ● | ● | | | | |
| Financial Costs | FinanCost | € | ● | ● | ● | | | | ● | ● | ● | | | | |
| Financial Needs | FinanNeed | € | | ● | ● | | | | ● | | ● | | | | |
| Risk solvency | RiskSolv | % | | ● | ● | | | | ● | | ● | | | | |
| Annualized ROI | AnnROI | % | | ● | ● | | ● | | ● | ● | ● | | | | ● |
| Entities involved | #Ent | no. | ● | ● | ● | ● | | ● | ● | ● | ● | ● | ● | | |
| Cost renovation target A | PECese | €/Entity<br>€/Community | | ● | ● | ● | ● | ● | | ● | | ● | ● | | |
| Cost renovation target B | PECacc | €/Entity<br>€/Community | | ● | ● | ● | ● | ● | | ● | | ● | ● | ● | |
| Cost renovation target C | PECkWpx | €/Entity<br>€/Community | | ● | ● | ● | ● | ● | | ● | | ● | ● | | |
| Manag. & maint. costs (Equip.) | Opee | €/Building<br>€/Community | | ● | ● | ● | | | | ● | | ● | ● | | |
| Manag. & maint. costs (RES inst.) | Opex | €/Building<br>€/Community | | ● | ● | ● | | | | ● | | ● | ● | | |
| Primary energy demand reduction | Pedx | % | | ● | | ● | | ● | | | | ● | | | ● |
| Energy balance (Primary energy) | Bex | ΣkWh montly<br>ΣkWh yearly | | ● | | ● | ● | ● | | | | | | ● | ● |
| GhG emissions reduction | Emx | % TnCO2 yearly | | ● | | ● | | ● | | | | ● | | | ● |
| EPC implementation (upgrade) | EPCx | Value in range scale | | ● | | ● | ● | | | | ● | | | ● | ● |
| Installable peak power | PVpx | kWp | ● | ● | ● | ● | | | ● | ● | | ● | ● | | ● |
| Energy produced per Entity | PVex | kWh hourly<br>ΣkWh yearly | ● | ● | | ● | ● | | ● | ● | | | ● | | ● |
| Hourly balances per Ent. (Exp.or Surplus) | Rox | kWh hourly | ● | ● | | ● | | ● | | ● | | | ● | | ● |

Table 3. Cont.

| Outputs (KPIs) | Acronym | Units | Phase | | | | Stakeholder | | | | | Scope | | | |
|---|---|---|---|---|---|---|---|---|---|---|---|---|---|---|---|
| | | | Leading phase | Demand aggregation | Concept & Techn. design | Construction & Use | End-users | Public Authorities | Financial institutions | Property developers | Financial Appraisal | Renovation Strategy | Energy Community | Welfare & Security | Marketing |
| Hourly balances per Com. (Exp.) | Roc | kWh hourly ΣkWh yearly | ● | ● | | ● | | ● | | ● | | | ● | | ● |
| Electrical storage capacity | EE | kW | | ● | | ● | ● | | | | | ● | ● | | |
| Savings on the energy bill | EFx | €/Yearly entity | | ● | | ● | ● | | | | | ● | | ● | ● |
| Potential RES installation subsidies | Subx | €/kWp unit €/kWp total | ● | ● | | ● | ● | | | | ● | ● | ● | ● | ● |
| Value Tax (IBI) bonus | Imp | €/Yearly entity | ● | ● | | ● | ● | ● | | | | ● | ● | ● | ● |
| Value Tax (IBI) bonus for RES inst. | Δimp | % | ● | ● | | ● | ● | ● | | | | ● | ● | | ● |
| Total yearly bonus | Σimp | €/Yearly | ● | ● | | ● | | ● | | | | ● | ● | ● | ● |
| Payback period | Pbck | Years | ● | ● | | ● | ● | | ● | ● | ● | ● | ● | ● | ● |
| Affordability of energy | Afenx | %/Population | | ● | | ● | ● | ● | ● | | | | ● | ● | ● |
| Affordability of housing | Afhox | %/Population | | ● | | ● | ● | ● | ● | ● | | ● | ● | ● | ● |
| Personal safety | Safex | Value in range scale | | ● | | ● | ● | ● | | | | ● | | ● | ● |
| Energy consciousness | Encos | Value in range scale | | ● | | ● | | ● | | | | | ● | ● | ● |
| Healthy community | HeCom | %/Population | | ● | | ● | ● | ● | | | | | | ● | ● |
| Employment opportunities | Jobx | no./yearly | | ● | | ● | | ● | ● | ● | ● | | | | ● |

The stakeholders will be different beneficiaries of the simulated KPIs according to each phase of the renovation process as reported in Table 3. Due to the analysis of the current state and the model's accuracy, stakeholders are allowed to influence the model's computation phase by selecting and modifying the desired settings for both the calculation and data, by providing detailed input. Moreover, according to a wide PEDRERA approach beyond a reliable business tool, the model is meant to offer additional services, empowered by its next implementation into the urbanZEB tool web platform. For this reason, it is necessary to first establish each participant stakeholder and their characteristics clearly in order to effectively define the model results.

With regard to the **financial appraisal**, the design of the model algorithms has been concluded and tested on real large-scale renovation processes (i.e., the "ACR Pirineus" intervention on 32 buildings in Santa Coloma-Barcelona) [35]. In addition to its validation, the sensitivity analysis of the financial appraisal model was conducted (Figure 8) on 200 scenarios in order to determine how target variables (KPIs) are affected based on changes in other input variables. This simulation analysis refers also to show the outcome of a decision given a certain range of variables of the inputs (e.g., the execution budget (PEM), the number of entities involved in the process, the percentage of grants and of defaulters, the operational cost from the private partner or the financial costs in a shifting PPP model) in order to evaluate, for example, the revolving funds impact on the Public Administration or the payment rates covered by each user type, among others. Since various private and public financial mechanisms for energy renovations in buildings are currently available, the financial appraisal can range from well-established and traditional mechanisms such as grants, subsidies and loans to emerging new models and other oriented PPP models. The test beds demonstrated that, although public bodies are typically involved in large-scale retrofitting projects, the majority of them are only partially engaged, often playing a role in the subsidy plan or, occasionally, by allowing the legal framework to adapt to local conditions. At the same time, several end-user typologies need to be identified in order to better assess different scenarios and apply the most suitable solutions suited on their profile. For this scope five types of users have been stated according to the number of fees and the duration of the loan (between 5 and 10 years) (Table 4). Due the economic vulnerability of some low-income users to cover the cost of the renovation effort, for this reason a specific grant program has been identified for eligible homeowners (vulnerable persons) who are unable to pay their fees. It consists of the registration of a charge in the Property Register equivalent to the overall fee to be paid. This charge must be paid to the City Council in that case of the transfer of ownership (sale, inheritance or other). Thus, the financial capacity of the Public Sector determines the acceptable limit of the revolving funds.

**Table 4.** List of user types (UT) adopted in the financial appraisal (Figure 8). Source: authors.

| User Type (UT) | Description | Loan Duration (Years) | Number of Payments |
|---|---|---|---|
| UT 1 | Mode 2 scheduled payments | 0 | 2 |
| UT 2 | Mode 60 scheduled payments | 5 | 60 |
| UT 3 | Mode 120 scheduled payments | 10 | 120 |
| UT 4 | Inscription (for vulnerable persons) | 0 | 2 |
| UT 5 | Mode 96 scheduled payments | 8 | 96 |

Figure 8 presents the analysis considering two main KPIs related to the end-user perspective (average monthly payments) and the impact of revolving funds size on the Public Sector to cover vulnerable users (UT4). The graphics shows how the revolving funds will surpass 250 K€ when 3 different scenarios occur: (a) grant is lower than 15% of the total cost; (b) investment is high and the number of entities involved is considerable (more than 250 dwellings); (c) the proportion of UT4 increases from 10% to 20%. The amount of 250 K€ represents the suitable value adopted in the testbed and it is adjustable according to the effort that each Public Sector can assume to support limited number of parallel operations of the renovation process. The baseline scenario shown in Figure 8 represents the end-user and the Public Sector perspective when the conditions of Table 5 occur, referring to the Santa Coloma test bed.

The conclusions of the sensitivity analysis for the economic model demonstrate the model is robust enough to allow for different breakdowns between user types, variations in operational costs, variations in financial costs (i.e., interest rates), investment per dwelling and number of entities involved. In those cases, robustness refers to whether final monthly end-user payments remain lower than 100€ and savings offer incentive to undergo a large-scale retrofitting operation. Moreover, large operations with a high number of entities (i.e., 500) or more vulnerable users that may require access to municipality grants do increase both financial need and municipal resources in terms of operational cost and size of revolving fund. In such cases, the size of the operation can be a limiting factor for the Public Sector. Furthermore, the debt financing in the form of loans represents a more sustainable means of up-scaling energy efficiency investments as loans can provide liquidity and direct access to capital, as well as support the cashflow during the process period. Loans can be more relevant for energy efficiency measures attached to high upfront costs, especially in deep renovation projects which comprise a package of multiple intervention measures. Despite this, the market interest rate (TAE) deviations are included in the model, and the result of the financial appraisal reveals how the fluctuance of interests has a strong impact on the financial cost during the loan period, and how the business model will be consequently affected by the financial cost increase.

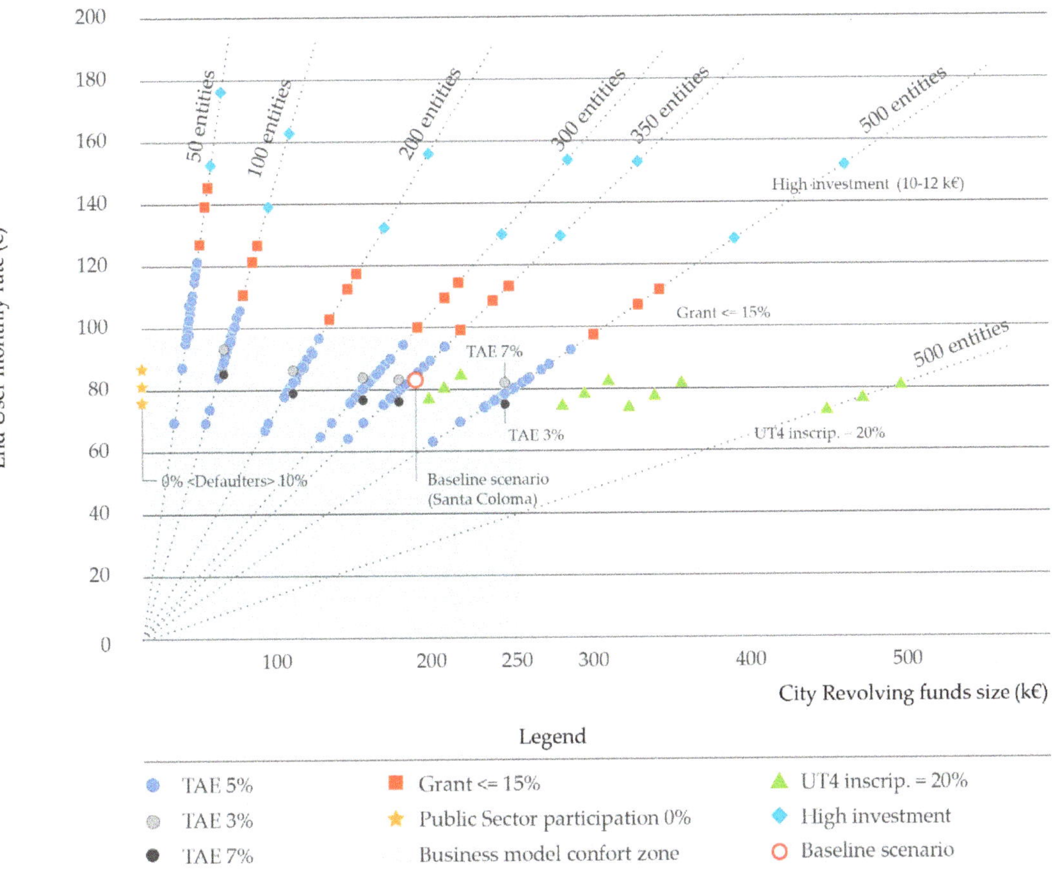


**Figure 8.** Extract of the PEDRERA sensitive analysis on the financial appraisal KPIs: the end-user perspective (UT2 monthly rate fee) and the Public Sector perspective (City revolving funds size) to cover vulnerable type users.

**Table 5.** List of input from Santa Coloma test bed, adopted as a baseline in the financial appraisal (Figure 8). Source: authors.

| Parameter | Definition | Acronym | Unit | Values |
|---|---|---|---|---|
| Entry data | # Entities | #Ent | no. | 350 |
| | Premises (% of Locals) | Prem%_Inc | % | 10% |
| PEM | Renovation investment (PEM/Ent) | PEM€_Ent | € | 5800 |
| | Increment Exec. Works | Work%_Inc | % | 0 |
| Technical project | Technical Project fees | Tech%_Fee | % | 13% |
| | Public Grant (covered by Public Sector) | Grant% | % | 35% |
| | Operational costs Fix (50 entities) | Opex€_Fix | € | 75,000 |
| | Operational costs Var. | Opex€_Var | € | 385 |
| | Corner Stone OverHead | OvH%(Cs) | % | 40% |
| | Corner Stone Benefit | Ben%(Cs) | % | 40% |

**Table 5.** *Cont.*

| Parameter | Definition | Acronym | Unit | Values |
|---|---|---|---|---|
| Financial Cost | Yearly TAE | TAE%(i) | % | 5% |
| | VAT Technical and Opex (k = 2, 3) | VAT% (k2,3) | % | 21% |
| | VAT Construction (k = 1) | VAT% (k1) | % | 10% |
| Economies of Scale | Incr. Market value after ref. | Market%_Inc | % | 20% |
| | Reduction on project fees | Proj%_Red | % | 15% |
| | Reduction on contract (tender offer) | Contr%_Red | % | 15% |
| | Reduction on PEM (market prize) | PEM%_Red | % | 10% |
| OPEX Breakdown | Phases (Fraction Step) | F_Step % | % | 10%, 30%, 20%, 15%, 25% |
| | PS (Public Sector) | PPP_PS% | % | 100%, 30%, 30%, 20%, 30% |
| | CS (Corner Stone) | PPP_CS% | % | 0%, 70%, 70%, 80, 70% |
| UTs (User Types) | Mode 2 scheduled payments | UTA (1) | % | 10% |
| | Mode 60 scheduled payments | UT A (2) | % | 70% |
| | Mode 120 scheduled payments | UT A (3) | % | 10% |
| | Inscription (for vulnerable persons) | UT B (4) | % | 10% |
| | Other | UTA (5) | % | 0% |
| | Defaulters | UTC (6) | % | 5% |

## 4. Discussions and Conclusions

In this paper, we presented a description of the approach, purpose, methods and first results obtained by the ongoing research project PEDRERA focused on the design of a PEDs oriented renovation model. The input and KPIs considered for the sensitive analysis shown in the paper (Tables 1 and 3, and Figure 8) mean the main aspects covered by the PEDRERA model reach each "scope", according to a wider PED vision where both energy efficiency and production strategies are considered together with the operative, social, economic and financial aspects.

The PEDRERA model is currently being developed with Python, a well-known programming language, the same as the urbanZEB tool. Once the programming phase will be completed then the software will be fully implemented in the web platform, thus delivering the multiple stakeholders' engagement in large-scale renovation actions, and will support the prediction of the sustainability and positive outcomes of distinct renovation scenarios.

Other areas of development will go in the direction of the Renovation Wave, with different solutions and services that can help to face challenges in terms of supporting renovation program and seeking for innovative financial frameworks. With this perspective, the PEDRERA model implementation will provide a very comprehensive service able to support and promote renovation actions by multidimensional and dynamic scenarios analysis, as well as the prediction of the potential impacts and benefits from feasible measures both at building and district level. In our future work, we plan to further develop a web platform tool service from the PEDRERA model, in order to boost and design large-scale renovation actions as well as engage the different stakeholders in the renovation process. Further steps will be to adopt the platform prototype in ongoing EU-wide efficient retrofitting projects at district level. Specific case studies will be selected to ensure the platform performance is tested under different conditions including climate aspects, boundary conditions, uses, building typologies, intervention levels, conservation conditions and other aspects.

Coherently to the deployment of the Clean Energy Transition and the Driving Urban Transition (relevant to the Renovation Wave), the PEDRERA model already establishes appropriate mechanisms to analyze the interdisciplinary aspects addressed by the energy communities (ECs). This can only be done by leveraging a range of very advanced analysis including urban modelling and interoperability of data as well as information from a spread digitalization of cities. According to this pathway, semantic frameworks will help to spatialize, organize and normalize information, as also making possible the graphic representation, the insertion of algorithmic-logical models as well as the realization of complex questions. Nevertheless, there is still a lack of tools/services oriented to the relevant key stakeholders of the building renovation process. Thus, one of the main contributions of the project regards leading and delivering innovation of energy saving and renewables-related services but also improving the consciousness on renovation projects opportunities from the very beginning in order to force and support large-scale actions.

Given to today's pressing scenarios of energy transition and climate change, and to the economic worldwide circumstances worsened by the COVID-19 pandemic, the energy poverty is one of the main issues that will profoundly characterize urban environments in the next years. The lack of capital is clearly one of the most pressing issues: although many large-scale private funds are eager to find and finance bankable projects, nevertheless the fragmented nature of the renovation market and actors (at least until solutions to deliver high volumes of renovation are available) hinders their interest and ability to fund building renovation at a large scale. In order to take informed decisions in their respective realms, these stakeholders need to have access to information which suits their knowledge and capacities. Furthermore, the possibilities of using city model maps are inherently unlimited and can be addressed to current and future city issues that may arise.

**Author Contributions:** Conceptualization, P.C., J.P., J.S.; methodology, P.C., J.P., J.S., J.A.A. and A.B.F.; formal analysis, P.C., J.P., J.S., J.A.A. and A.B.F.; investigation, P.C., J.P., J.S., J.A.A. and A.B.F.; writing—original draft preparation, P.C., J.P., J.S., J.A.A. and A.B.F.; writing—review and editing, P.C.; visualization, P.C. and A.B.F.; supervision, P.C., J.P., J.S. All authors have read and agreed to the published version of the manuscript.

**Funding:** The research leading to these results has received funding from the European Union's Horizon 2020 research and innovation program under the Marie Skłodowska-Curie grant agreement No 712949 (TECNIOspring PLUS) and from the Agency for Business Competitiveness of the Government of Catalonia. TECNIOspring PLUS. Investigator: Paolo Civiero, Project: PEDRERA. Positive Energy Districts renovation model. IREC (Barcelona—ES).

**Institutional Review Board Statement:** Not applicable.

**Informed Consent Statement:** Not applicable.

**Data Availability Statement:** Not applicable.

**Conflicts of Interest:** The authors declare no conflict of interest.

# References

1. Clerici Maestosi, P.; Andreucci, M.B.; Civiero, P. Sustainable Urban Areas for 2030 in a Post-COVID-19 Scenario: Focus on Innovative Research and Funding Frameworks to Boost Transition towards 100 Positive Energy Districts and 100 Climate-Neutral Cities. *Energies* **2021**, *14*, 216. [CrossRef]
2. MITMA. ERESEE 2020. Actualización 2020 de la Estrategia a Largo Plazo para la Rehabilitación Energética en el Sector de la Edificación en España. June 2000. Available online: https://www.mitma.gob.es/recursos_mfom/paginabasica/recursos/es_ltrs_2020.pdf (accessed on 20 April 2021).
3. EC. A Renovation Wave for Europe—Greening Our Buildings, Creating Jobs, Improving Lives, COM (2020) 662. Available online: https://eur-lex.europa.eu/legal-content/EN/TXT/?qid=1603122220757&uri=CELEX:52020DC0662 (accessed on 20 April 2021).
4. Casals, X.G.; Sanmartí, M.; Salom, J. *Smart Energy Communities*; ICAEN: Madrid, Spain, 2019; pp. 1–189. Available online: http://icaen.gencat.cat/web/.content/10_ICAEN/17_publicacions_informes/11_altres_publicacions/arxius/SmartEnergyCommunities.pdf (accessed on 20 April 2021).
5. JPI UE. PED Programme Management. Europe towards Positive Energy Districts. Booklet. Available online: https://jpi-urbaneurope.eu/wp-content/uploads/2020/06/PED-Booklet-Update-Feb-2020_2.pdf (accessed on 20 April 2021).

6. Civiero, P.; Salom, J.; Pascual, J. Aggregated data management and business model in designing Positive Energy Districts. In Proceedings of the EAAE—ARCC International Conference & 2nd Valencia International Biennial of Research in Architecture: The Architect and the City: Vol. 1, Valencia, Spain, 11–14 November 2020; pp. 908–917, ISBN 978-84-9048-981-9. Available online: https://www.eaae-arcc-ic.upv.es/files/2021/02/EAAE-ARCC-IC-2nd-VIBRArch_The-Architect-and-the-city_Volume-1.pdf (accessed on 12 May 2021).

7. Ferrando, M.; Causone, F.; Hong, T.; Chenc, Y. Urban building energy modeling (UBEM) tools: A state-of-the-art review of bottom-up physics-based approaches. *Sustain. Cities Soc.* **2020**, *62*, 102408. [CrossRef]

8. Civiero, P.; Sanmartí, M.; García, R.; Gabaldón, A.; Adrés Chicote, M.; Ferrer, J.A.; Ricart, J.E.; Franca, P.; Escobar, G.J. Distritos de Energía Positiva (PEDs) en España, Una propuesta de Iniciativa Tecnológica Prioritaria de la PTE-EE. In Proceedings of the VII Congreso EECN, Libro de Comunicaciones y Proyectos, Madrid, Spain, 5 November 2020; Grupo Tecma Red, S.L.: Madrid, Spain, 2020. Available online: https://www.construible.es/biblioteca/libro-comunicaciones-proyectos-7-congreso-edificios-energia-casi-nula (accessed on 20 April 2021).

9. Cocchia, A. Smart and digital city: A systematic literature review. *Smart City* **2014**, 13–43. [CrossRef]

10. EC. Consolidated text: Directive 2007/2/EC of the European Parliament and of the Council of 14 March 2007 Establishing an Infrastructure for Spatial Information in the European Community (INSPIRE). Available online: http://data.europa.eu/eli/dir/2007/2/2019-06-26 (accessed on 20 April 2021).

11. EU. Commission Implementing Decision 2019/1372 of 19 August 2019 implementing Directive 2007/2/EC of the European Parliament and of the Council as Regards Monitoring and Reporting (Notified under Document C(2019) 6026). Available online: http://data.europa.eu/eli/dec_impl/2019/1372/oj (accessed on 20 April 2021).

12. Arcas Abella, J.; Pages-Ramon, A. UrbanZEB, Estrategias urbanas de transición energética de edificios. In Proceedings of the 14th National Congress Conama 2018, Vigo, Spain, 26–29 November 2020; Fundación Conama: Vigo, Spain, 2020. Available online: http://www.conama11.vsf.es/conama10/download/files/conama2018/CT%202018/222224182.pdf (accessed on 20 April 2021).

13. Arcas Abella, J.; Pagès Ramon, A.; Bilbao, A. urbanZEB: Estrategias urbanas de transición energética de edificios. In Proceedings of the IV Congreso ISUF-H: Metrópolis en Recomposición: Prospectivas Proyectuales en el Siglo XXI: Forma urbis y Territorios Metropolitanos, Barcelona, Spain, 28–30 September 2020; DUOT, UPC: Spain, 2020; pp. 1–19, ISBN 978-84-9880-841-4. Available online: http://hdl.handle.net/2117/328899 (accessed on 20 April 2021).

14. Reinhart, C.F.; Davila, C.C. Urban building energy modeling—A review of a nascent field. *Build. Environ.* **2016**, *97*, 196–202. [CrossRef]

15. Cerezo Davila, C.; Reinhart, C.F.; Bemis, J.L. Modeling Boston: A workflow for the efficient generation and maintenance of urban building energy models from existing geospatial datasets. *Energy* **2016**, *117*, 237–250. [CrossRef]

16. Ang, Y.Q.; Berzolla, Z.M.; Reinhart, C.F. From concept to application: A review of use cases in urban building energy modeling. *Appl. Energy* **2020**, *279*, 115738. [CrossRef]

17. Yanwen, L.; Jiang, H.; Yuting, H. A rule-based city modeling method for supporting district protective planning. *Sustain. Cities Soc.* **2017**, *28*, 277–286. [CrossRef]

18. Nouvel, R.; Mastrucci, A.; Leopold, U.; Baume, O.; Coors, V.; Eicker, U. Combining GISbased statistical and engineering urban heat consumption models: Towards a new framework for multi-scale policy support. *Energy Build* **2015**, *107*, 204–212. [CrossRef]

19. Sola, A.; Corchero, C.; Salom, J.; Sanmarti, M. Simulation Tools to Build Urban-Scale Energy Models: A Review. *Energies* **2018**, *11*, 3269. [CrossRef]

20. Monteiro, C.S.; Costa, C.; Pina, A.; Santos, M.Y.; Ferrão, P. An urban building database (UBD) supporting a smart city information system. *Energy Build.* **2018**, *158*, 244–260. [CrossRef]

21. Ascione, F.; De Masi, R.F.; de'Rossi, F.; Fistola, R.; Vanoli, G. Energy Assessment in Town Planning: Urban Energy Maps. *WIT Trans. Ecol. Environ.* **2012**, *155*, 205–216. [CrossRef]

22. Cupelli, L.; Schumacher, M.; Monti, A.; Mueller, D.; De Tommasi, L.; Kouramas, K. Simulation Tools and Optimization Algorithms for Efficient Energy Management in Neighborhoods. In *Energy Positive Neighborhoods and Smart Energy Districts*; Academic Press: Cambridge, MA, USA, 2017; pp. 57–100.

23. García-Fuentes, M.A.; Hernández, G.; Serna, V.; Martín, S.; Álvarez, S.; Lilis, G.N.; De Tommasi, L. OptEEmAL: Decision-Support Tool for the Design of Energy Retrofitting Projects at District Level. In *CESB19. IOP Conference Series: Earth and Environmental Science*; IOP Publishing: Bristowl, UK, 2019; Volume 290, p. 012129. [CrossRef]

24. OptEEmAL. The Solution for Designing Your Energy Efficient District Retrofitting Project; OptEEmAL Final Booklet; Steinbeis-Europa-Zentrum der: Germany. 2019. Available online: https://www.opteemal-project.eu/files/opteemal_final_booklet_web.pdf (accessed on 8 May 2021).

25. Madrazo, L. *The Social Construction of a Neighbour-Hood Identity*; Open House International: Eindhoven, The Netherlands, 2019; Volume 44, pp. 71–80. [CrossRef]

26. Visscher, H.; Dascalaki, E.; Sartori, I. Towards an energy efficient European housing stock: Monitoring, mapping and modelling retrofitting processes. *Energy Build.* **2016**, *132*, 1–154. [CrossRef]

27. Civiero, P. *Tecnologie per la Riqualificazione. Soluzioni e Strategie per la Trasformazione Intelligente del Comparto Abitativo Esistente*, 1st ed.; Maggioli Editore: Santarcangelo di Romangna, Italy, 2017; pp. 1–240. [CrossRef]

28. Ortiz, J.; Casquero-Modrego, N.; Salom, J. Health and related economic effects of residential energy retrofitting in Spain. *Energy Policy* **2019**, *130*, 375–388. [CrossRef]

29. Pasichnyi, O.; Wallin, J.; Kordas, O. Data-driven building archetypes for urban building energy modelling. *Energy* **2019**, *181*, 360–377. [CrossRef]
30. TABULA. Typology Approach for Building Stock Energy Assessment, TABULA Project (2009–2012). Available online: https://episcope.eu/ (accessed on 10 May 2021).
31. Cuchí, A.; Sweatman, P. *Informe GTR 2011, Una Visión-País para el Sector de la Edificación en España, Hoja de ruta para un Nuevo sector de la Vivienda*; Fundación Conama: Madrid, Spain, Online Version; 2011; pp. 1–70. Available online: http://www.encuentrolocal.vsf.es/download/bancorecursos/libro_GTR_cast_postimprenta.pdf (accessed on 20 April 2021).
32. Ortiz, J.; Guarino, F.; Salom, J.; Corchero, C.; Cellura, M. Stochastic model for electrical loads in Mediterranean residential buildings: Validation and applications. *Energy Build.* **2014**, *80*, 23–36. [CrossRef]
33. IDESCAT. Index Socioeconòmic Territorial. Available online: https://www.idescat.cat/dades/ist/mapes/ (accessed on 8 May 2021).
34. Salom, J.; Pascual, J. *Residential Retrofits at District Scale. Business Models under Public Private Partnerships*; InnoEnergy: Eindhoven, The Netherlands, Online; 2018; pp. 1–75. ISBN 978-84-09-07914-8. Available online: https://www.buildup.eu/en/node/57005 (accessed on 20 April 2021).
35. Ayuntamiento de Santa Coloma de Gramenet. Renovem els Barris. Available online: https://www.gramenet.cat/ajuntament/arees-municipals/renovem-els-barris/ (accessed on 20 April 2021).

*Article*

# Possibilities of Upgrading Warsaw Existing Residential Area to Status of Positive Energy Districts

**Hanna Jędrzejuk * and Dorota Chwieduk**

Institute of Heat Engineering, Faculty of Power and Aeronautical Engineering, Warsaw University of Technology, 00-665 Warsaw, Poland; Dorota.Chwieduk@itc.pw.edu.pl
*    Correspondence: Hanna.Jedrzejuk@pw.edu.pl

**Abstract:** This paper analyses possibilities of refurbishment of Warsaw's residential buildings towards standards of the Positive Energy District. The annual final energy consumption in the city in 2019 for the district heating was 8668 GWh, gas (pipelines) was 5300 GWh, electricity from the grid was 7500 GWh, while the emission of the carbon dioxide was $5.62 \times 10^9$ kg. The city consists of 18 districts, which are heterogeneous in terms of typology and structure of buildings. The great variety of buildings can be seen, for example, by the annual final energy demand for space heating and hot water preparation per unit of room area. This annual index ranges from over 400 kWh/m$^2$ in historic buildings to 60 kWh/m$^2$ in modern buildings. A reduction in the consumption of non-renewable energy sources and carbon dioxide emissions can be achieved by improving the energy standard of residential buildings and by using renewable energy sources: solar energy, geothermal energy and biogas. The potential barriers for achieving the status of a positive energy district, for example, problems connected with ownership, financing new investments and refurbishment and legal boundaries, have been identified. Moreover, changing the existing electrical grid and district heating systems in urban areas in Warsaw requires comprehensive modernization of practically the entire city's infrastructure.

**Keywords:** positive energy districts; residential buildings; district heating; renewable energy resources

**Citation:** Jędrzejuk, H.; Chwieduk, D. Possibilities of Upgrading Warsaw Existing Residential Area to Status of Positive Energy Districts. *Energies* **2021**, *14*, 5984. https://doi.org/10.3390/en14185984

Academic Editor: Paola Clerici Maestosi

Received: 14 July 2021
Accepted: 15 September 2021
Published: 21 September 2021



## 1. Introduction

The first intensive measures aimed at improving energy security appeared in response to the fuel crisis of 1973 [1]. Indirectly, this global shock also contributed to increasing the importance of environmental protection to preserve the environment for future generations. There are several standard actions aimed at reducing the burden of harmful substances in the environment: rationalization of needs, improvement of broadly understood efficiency, and increased use of "clean" technologies to meet the needs.

In the residential buildings sector, these activities are mentioned:
- rationalizing the behaviour of building users (residents),
- rationalizing the internal systems operation,
- reduction of heat losses in buildings by improving the thermal insulation of the envelope,
- recovery of heat discharged to the environment, e.g., from ventilation air, domestic sewage,
- improving the energy efficiency of all technical devices and systems.

Nevertheless, it is not enough just to try to reduce the energy demand to meet the needs. Hence the idea of using "clean" technologies that do not burden the environment, especially renewable energy ones, have consequently led to the creation of "nearly zero-energy", "zero-energy" and "energy-positive" buildings in terms of final energy and/or primary energy. The assessment also concerns the balance of emissions of harmful solid and gaseous substances, e.g., carbon dioxide ($CO_2$).

In the European Union, these activities are directed, and at the same time supported, by the Directives. For example, Directive 2002/91/EC of the European Parliament and of the Council of 16 December 2002 on the energy performance of buildings [2] has brought lasting change in the Polish construction law and regulations relevant to the technical requirements of newly constructed buildings and buildings under refurbishment [3]. Nowadays, a comprehensive approach to energy conservation and environmental protection issues, linked to climate change problems, is extremely valuable and is evident in Polish national energy policy [4]. The energy and climate policies have been linked very strongly and utilisation of renewable energy technologies has become one of the basic elements of both.

New or renovated buildings are evaluated on the basis of energy consumption of non-renewable primary energy for space heating, ventilation, cooling, domestic hot water and lighting. It is also necessary to determine $CO_2$ emissions and the share of renewable energy sources in meeting the energy demand [3,5].

The City of Warsaw is actively engaged in activities aimed at rationalizing and reducing energy use. An example of completed actions include action plans for sustainable energy consumption for Warsaw by 2020 [6] and investigations on the development of the Warsaw housing standard [7]. Warsaw, in cooperation with the local authorities of Austria, Germany and Italy, took part in the "Cities on Power" project (2011–2014), with the aim to increase the use of renewable energy in urban areas. One of the results was estimation of the possibilities of using renewable energy sources in Warsaw [8]. Future planning has been performed for the 2050 perspective and is constantly updated [9]. In 2016, the City of Warsaw began participating in the Cities Council of the EERA (European Energy Research Associations) Smart Cities Joint Program.

In 2019, at the capital city of Warsaw, the Office of Air Protection and Climate Policy was established by integration of four different former Offices. The new Office's tasks include preparing and updating action plans in the field of air protection and climate change and elaborating reports on their implementation, including cooperation with competent authorities in the preparation of higher-level programs (regional, national) [10]. Other examples can be mentioned, e.g., the Municipal Climate Adaptation Plan [11], and the Low-Emission Economy Plan for the Capital City of Warsaw [12]. Among the numerous recommended actions, the introduction of solutions increasing the energy independence of the city, including increasing the share of energy from renewable sources, was mentioned. What is really important is that plan was the result of joint work of residents, entrepreneurs, representatives of various types of organizations and the City of Warsaw.

In Section 2, administrative characteristics of Warsaw are presented, and some historical facts are also mentioned. Because of the topic of the paper, the housing stock and energy sector are analysed with more attention. Considerations presented in Section 2 give the background for more detailed analysis of a very specific situation in terms of supplying heat to end users in Warsaw using a central district heating system. This analysis is carried out in Section 3. It can be noticed that with having such a well-developed district heating system, it is difficult to implement new investments in energy systems that use solutions other than connecting new buildings to the existing network. However, if new housing estates are built in an area that does not have access to the central energy systems, then the extension of the existing energy network may be technically too complicated and, consequently, too expensive. As a result, it is more reasonable to build new local energy systems in accordance with the current requirements of energy conservation and environmental protection. Consequently, this provides the basis for the use of energy from renewable sources available in the area of the city. Section 4 is therefore devoted to the presentation of the theoretical potential of using renewable energy sources in the city and its outskirts. Section 5 includes subchapters devoted to a different renewable energy, the theoretical potential of which in Warsaw or on its outskirts indicates the possibility of its use for utility purposes. However, possible and effective use can be limited due to specific location, environmental or legal aspects, which are also mentioned in Sections 5 and 6. Section 6 briefly presents what should be done in order to achieve the goals set for the Positive

Energy Districts (PED) in a city such as Warsaw. Moreover, it shows the potential risks that hinder the achievement of the goals. Finally, conclusions are formulated in Section 6.

## 2. Administrative Characteristics of Warsaw: Housing Stock and Energy Sector Description

The settlement in today's Warsaw dates back to the 9th or 10th century. However, the city was most likely founded at the turn of the 13th and 14th centuries, gaining its location around the year 1300. Warsaw became the capital of Poland in the year 1596, when the King Sigismund III Vasa decided to move the capital from Cracow to Warsaw [13]. Currently, there are approximately $1.8 \times 10^6$ inhabitants in Warsaw and the city covers an area of approximately $5.20 \times 10^8$ m$^2$. The average population density is approximately 3.46 people per square meter [14]. Administratively, Warsaw is divided into 18 districts (Figure 1) and Śródmieście is the historical and present city centre [15].

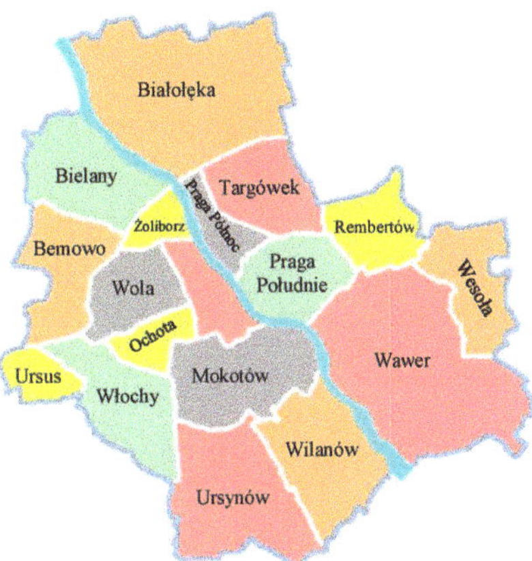


**Figure 1.** Administrative districts in Warsaw [15].

The largest area includes the following administrative districts: Wawer ($7.97 \times 10^7$ m$^2$), Białołęka ($7.00 \times 10^7$ m$^2$) and Ursynów ($4.38 \times 10^7$ m$^2$), while the smallest one is Żoliborz ($8.50 \times 10^6$ m$^2$) (Table 1). The largest residential area is in Białołęka ($1.00 \times 10^7$ m$^2$) (Table 1, [16]). The "greenest" districts are Wawer ($1.43 \times 10^7$ km$^2$) and Białołęka ($1.17 \times 10^7$ m$^2$). Additionally, taking into account the percentage of parks, lawns and green areas in housing estates in the district's area, Wawer (18%), Rembertów (17%) and Włochy (15%) can be distinguished in this regard. The districts with the largest share of transport routes are located in the outskirts of Warsaw, Wawer ($1.35 \times 10^7$ m$^2$, 17%) and Białołęka ($1.17 \times 10^7$ m$^2$, 15%) [17]. This is mainly due to the location of the main transport connections: internally within in the city and externally outside of the city.

The structure of the land by use is as follows: 28% residential areas, 28% green areas, 12% agricultural areas, 10% development and communication areas, 7% service areas, 3% surface water areas, 1% technical infrastructure areas and 6% other [18]. It can therefore be concluded that Warsaw has a relatively large biologically active area, which constitutes 42%.

The final energy consumption structure in the year 2014 [9], averaged for the long-term weather conditions, is shown in the Figure 2. Most important is the heat from the district heating network (41.87%) and electricity from the national grid (32.52%). The category "Others", which includes renewable energy systems, was still marginal (5.82%) that year.

**Table 1.** Area and population, total and by districts in the year 2020 as of 31 December [16].

| Districts | Area [m²] | Population [People] |
|---|---|---|
| Bemowo | 25,000,000 | 125,270 |
| Białołęka | 73,000,000 | 132,281 |
| Bielany | 32,300,000 | 130,848 |
| Mokotów | 35,400,000 | 217,424 |
| Ochota | 9,700,000 | 82,018 |
| Praga-Południe | 22,400,000 | 180,066 |
| Praga-Północ | 11,400,000 | 63,442 |
| Rembertów | 19,300,000 | 24,679 |
| Śródmieście | 15,600,000 | 111,338 |
| Targówek | 24,200,000 | 124,742 |
| Ursus | 9,400,000 | 62,399 |
| Ursynów | 43,800,000 | 151,288 |
| Wawer | 79,700,000 | 79,078 |
| Wesoła | 22,900,000 | 25,926 |
| Wilanów | 36,700,000 | 43,423 |
| Włochy | 28,600,000 | 44,343 |
| Wola | 19,300,000 | 142,694 |
| Żoliborz | 8,500,00 | 52,907 |
| Warsaw: TOTAL | 517,200,000 | 1,794,166 |

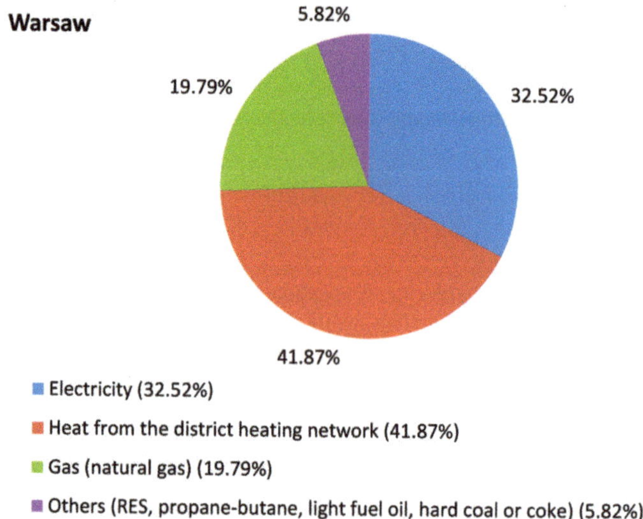


**Figure 2.** Final energy consumption structure in the year 2014 [9].

According to the data [9], buildings constructed in the period of 1945–1970 have the largest share of the total number of buildings in Warsaw (26.39%). The share of relatively new buildings constructed after 2002 accounts only for 4.60%.

The final as well as primary energy consumption indices depending on the construction period show how different buildings in terms of the energy efficiency are in Warsaw (Table 2) [9,19,20].

**Table 2.** Age structure of housing stock in Warsaw and energy consumption indices [9,19,20].

| Construction Period | Buildings | Final Energy Index [kWh/(m$^2$ a)] | Primary Energy Index [kWh/(m$^2$ a)] |
|---|---|---|---|
| Before 1918 | 12,200 | >300 | >350 |
| 1918–1944 | | 260–300 | 300–350 |
| 1945–1970 | 19,826 | 220–260 | 250–300 |
| 1971–1978 | 5691 | 190–220 | 210–250 |
| 1979–1988 | 7040 | 140–190 | 160–210 |
| 1989–2002 | 15,974 | 125–160 | 140–180 |
| 2003–2007 | 6464 | 90–120 | 100–150 |
| 2008–2011 | 4043 | 90–120 | 100–150 |
| TOTAL | 71,238 | | |

Low thermal energy buildings can be improved through the typical retrofit methods: reduction of thermal transmission of the building external partitions (e.g., windows, walls, roofs), elimination of the existence of thermal bridges, improvement of efficiency of heating systems as well as hot water preparation (e.g., exchange of heat source, introduction and optimization of energy control strategy), improvement of ventilation system and heat recovery from the exhausted air. The possible effects of the thermo-modernization in two variants, depending on the construction period and type of a building, of two variants are presented in the Table 3 [20].

**Table 3.** Reduction potential of the energy demand in houses [20].

| Construction Period | Single Family Houses | | Terraced Houses | | Multifamily Houses | |
|---|---|---|---|---|---|---|
| | Standard | Deep | Standard | Deep | Standard | Deep |
| up to 1945 | 63.2% | 72.7% | 55.7% | 71.4% | 57.8% | 71.9% |
| 1945–1966 | 61.8% | 72.2% | 30.8% | 53.8% | 52.7% | 64.9% |
| 1967–1985 | 53.4% | 62.8% | 18.2% | 47.7% | 51.6% | 64.4% |
| 1986–1992 | 41.9% | 50.0% | 21.1% | 42.1% | 36.5% | 53.8% |
| 1993–2002 | 34.4% | 43.8% | 21.9% | 39.4% | 26.3% | 46.9% |
| 2003–2008 | 25.7% | 35.7% | 21.4% | 35.7% | 21.4% | 39.3% |
| after 2008 | 29.1% | 38.3% | 22.2% | 34.8% | 21.4% | 39.3% |

The first variant "standard" means reaching the level of the reduction when current requirements are fulfilled [3], while "deep" means the application of more intensive technical upgrading actions, up to reaching the maximal possible effects assuming use of new but conventional technologies. It is interesting that the application of the current technical possibilities, even in buildings that have recently been put into use, allows for reducing the demand for final energy in residential buildings by approximately 35% to almost 40%. It is possible mainly due to the reduction of the heat demand for heating the ventilation air, e.g., replacing gravity ventilation with a mechanical ventilation system with heat recovery.

It is characteristic for old buildings that no changes in their external appearance are possible if they are under conservator's protection. Therefore, in such buildings, the thermal insulation of the external partitions can be added only from the inside of the building. The effectiveness of such treatments is low, and reduction of energy consumption does not exceed 20%. In many cases there is no possibility to install solar collector systems or PV systems due to heritage protection, so the old quarters of Warsaw are not so attractive in terms of generating energy on-site (locally). The most interesting in terms of retrofit are groups of buildings built in the 1970s and later, with the annual final energy consumption index for space heating and hot water preparing at a level of 220 kWh/m$^2$. In such cases the possible reduction of energy demand reaches 40%, but it is still too low for reaching positive energy building status.

Figure 3 presents maximum annual primary non-renewable energy indices for heating, ventilation and hot water preparation [3]. The strongest requirements are foreseen for educational buildings, while for residential buildings the requirements are less restrictive. Nevertheless, in all cases the limit values are constantly lowered, which helps to reduce the consumption of non-renewable primary energy and hence $CO_2$ emissions.

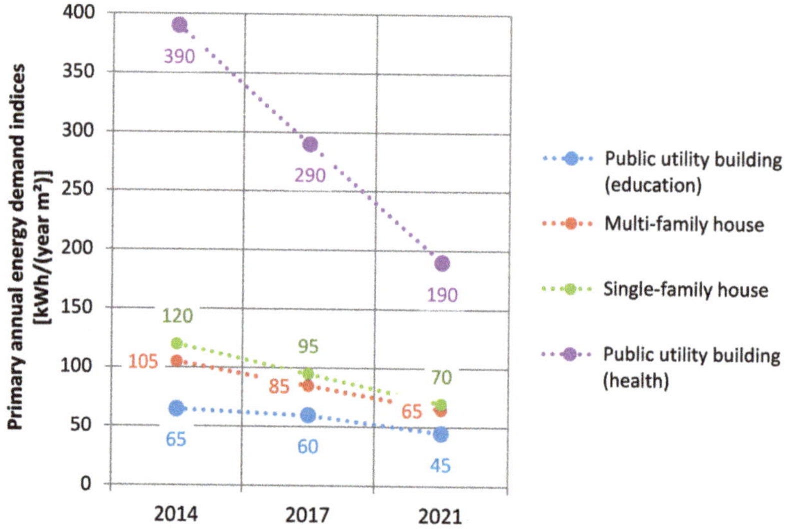


**Figure 3.** Maximum annual primary non-renewable energy indices for heating, ventilation and hot water preparation [3].

The annual final energy consumption in the year 2019 [9] for district heating was 8668 GWh, gas (pipelines) was 5300 GWh, electricity from the grid was 7500 GWh, while the emission of the carbon dioxide was $5.62 \times 10^9$ kg [9].

In the year 2019, the co-generation heat and power plants had a thermal capacity of 2245 MW [9]. Old coal-fired plants are expected to be decommissioned in the near future. They will be replaced by modern gas-steam units [21]. During peak demand, the system can use thermal energy of 1 GWh stored in a water tank [9].

The capacity of all electricity generation sources in the year 2019 was 8.529 GW, while electricity consumption was 7500 GWh, of which households consumed 2141 GWh while the electricity storage in the electricity grid was 2 MWh [9].

In the year 2014 [9], taking into account the type of heating source (system and fuel), the situation was as follows: district heating supplying heat to buildings of 80,789,204 m$^2$ of heated area (71.0%) and gaseous fuel for 23,395,279 m$^2$ (20.6%). There were other energy carriers used in small amounts: electrically driven heaters for 2,599,475 m$^2$ (2.3%), propane-butane mixture for 1,316,459 m$^2$ (1.2%), light fuel oil for 3,432,031 m$^2$ (3.0%), hard coal for 1,843,420 m$^2$ (1.6%) and renewable energy sources for 411,743 m$^2$ (0.4%).

Possibilities of the reduction of final energy consumption in residential buildings has been estimated by assuming average efficiency of the heating systems for all types of residential buildings in the specific period of their construction due to lack of detailed data. Additionally, it was assumed that historical buildings would not be refurbished. As the final energy reduction was decreased by almost 30% in the standard and 40% in the deep retrofit, relevant emissions in the residential sector are possible.

A characteristic feature of the demand for electric power is the lower demand in summer, but due to the increase in cooling demand, these relations will change (Figure 4). Although the available capacity of renewable energy sources is constantly increasing, it is still lower than 16% of the actual needs.

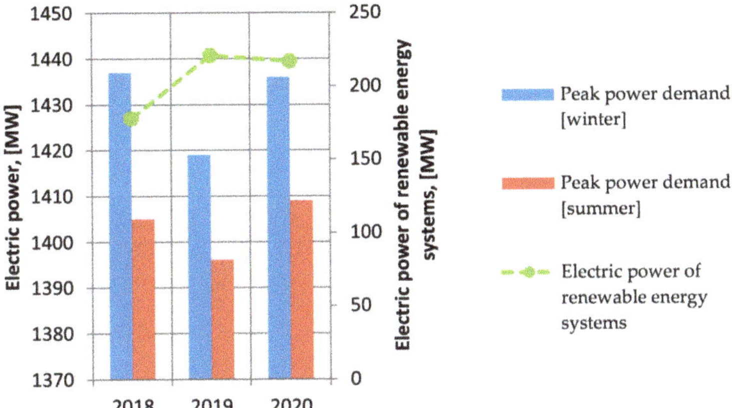



**Figure 4.** Electric power in Warsaw, [MW] [9].

Electricity consumption in the last three years has remained stable despite the continuous development of the city (Figure 5). Energy consumption in households in the following years has been similar and accounts for less than 29% of total energy consumption in the city. Unfortunately, the coverage of energy demand from renewable energy systems is less than 2%. However, the real share of renewable energy sources is higher because the official data do not take into account the energy produced by prosumers for their own needs [9].

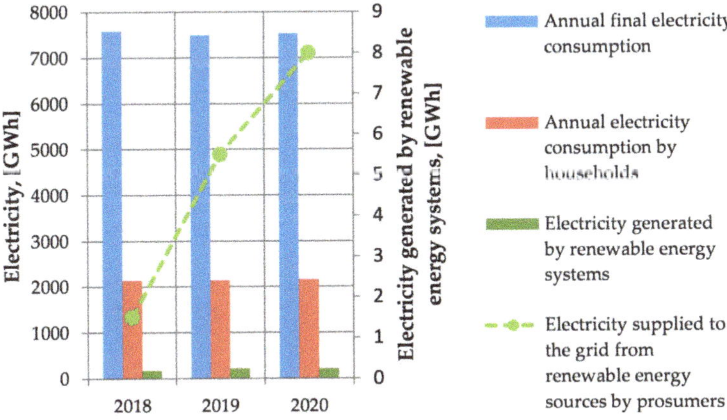



**Figure 5.** Electricity generation in Warsaw, [GWh] [9].

The basic rate of thermal energy necessary to cover the needs in Warsaw is generated in heat and power plants and heating plants (Figure 6). The available thermal installed capacity of renewable energy systems accounts for about 6% of the total demand.

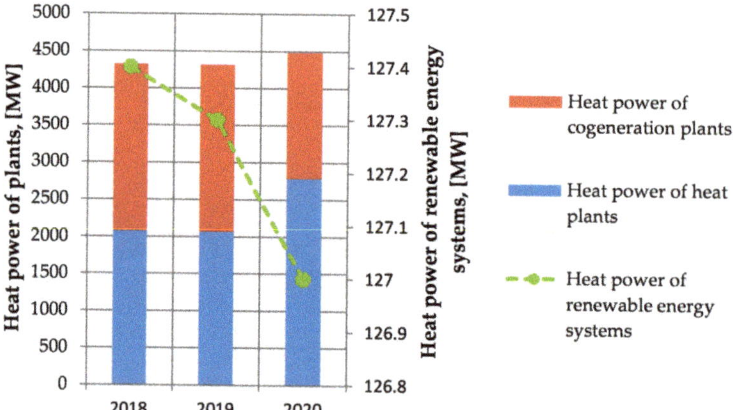

**Figure 6.** Thermal installed capacity of energy plants in Warsaw, [MW] [9].

The heat generated in renewable energy systems (Figure 7) accounts for less than 5% of the heat supplied to end users by the district heating network. However, it should be noted that the heat generated by individual users for their own needs is not included in the balance.

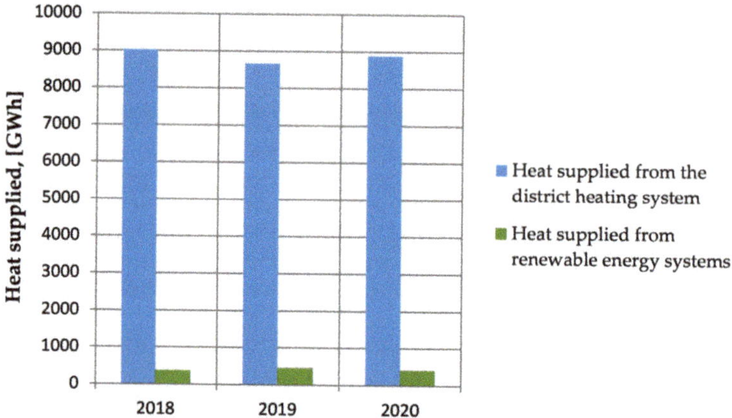

**Figure 7.** Heat supplied to the end users from the district heating network and renewable energy systems in Warsaw, [GWh] [9].

## 3. District Heating Network

The city of Warsaw is characterized by a very specific situation in terms of supplying heat to end users: the residential and public sector, services and industry. Heat is supplied centrally, ensuring cheap and reliable collection of this heat by users located in all parts of the city. Warsaw's district heating network is one of the largest in Europe, with a length of almost 1800 km. It supplies heat to over 80% of buildings in the city. Heat is produced in co-generation, and coal is the main energy fuel [22]. Polish district heating systems are high temperature systems since they can supply steam with temperatures of 120 °C. The present state and planned development of Warsaw's district heating system is presented in Figure 8 and it is mainly for residential purposes.

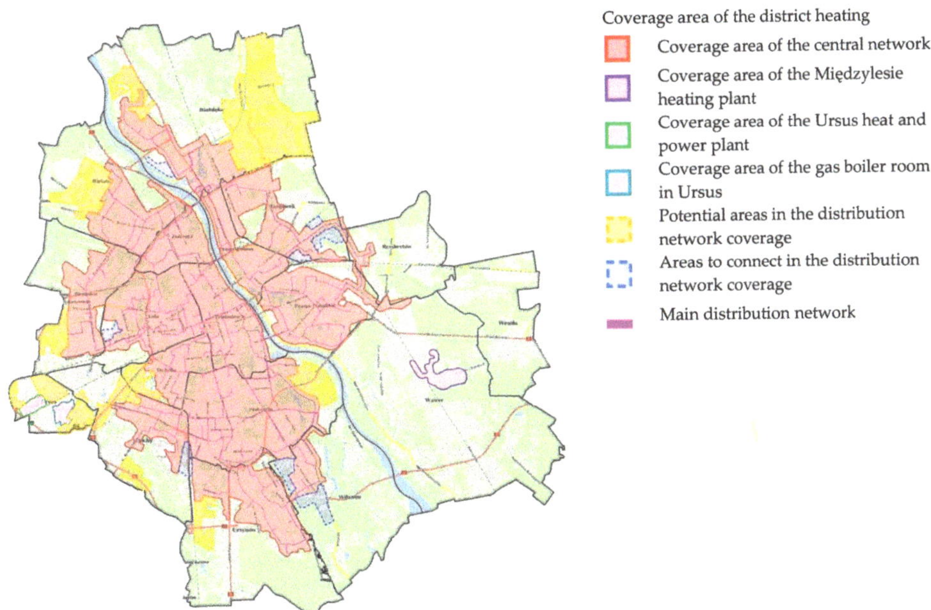



**Figure 8.** Coverage area of the district heating [8].

However, when analysing Figure 8, a question can be asked: why should the new residential districts (in yellow) be connected to the central district heating network? Extending the old network of poor quality (poor thermal insulation, corrosion of pipelines, leaks of heating medium (water or steam)) to new urban areas raises concerns not only in terms of energy efficiency, but also of the economic efficiency of such projects. The construction of new distributed energy systems focused on the use of own local energy resources, in particular renewable energies, seems to be a much more rational solution of energy and economic efficiency. Moreover, the construction of new buildings in newly emerging districts with the use of modern building materials and technologies enables a significant reduction in energy requirements for heating buildings. Newly created housing estates are usually characterized by low indices of energy needs. If, moreover, modern installation technologies are used, then, as a consequence, the final energy consumption can be significantly reduced and can easily meet modern building energy codes. Thanks to the use of renewable energy, the consumption of primary energy from fossil fuels can decrease significantly or even not be used at all. When using their own local and renewable energy resources, new estates can not only cope with their own small energy load but can also share energy with districts in the neighbourhood, becoming Positive Energy Districts. The next section presents and analyses the potential for using renewable energy in Warsaw. The different subchapters refer to the different renewable energies that can be used.

## 4. Potential of Renewable Energy Utilization in Warsaw

### 4.1. Geothermal Energy

In Poland, geothermal energy potential is rather low. There are resources of low enthalpy geothermal water which can be applied for heating and balneology needs. There are thermal plants where geothermal waters are effectively applied for heating, however their utilization is combined with heat pumps and gas boilers.

Geothermal water resources in Warsaw have been assessed as average [23], i.e., at a depth of 1000 m the temperature reaches 30–35 °C, at a depth of 2000 m it exceeds 55 °C, and at a depth of 3000 m it reaches even 85 °C [23,24]. Economic profitability, understood as competition with the currently used technologies, requires temperatures of a resource (geothermal water) to be at least 65 °C at a depth of 2000 m. As the prices of "conventional" energy and heat increase, the competitiveness of this technology increases.

The districts of Wawer, Białołęka and Ursynów have the best opportunities to use geothermal energy for heating buildings (Figure 9), but so far geothermal energy is not used.

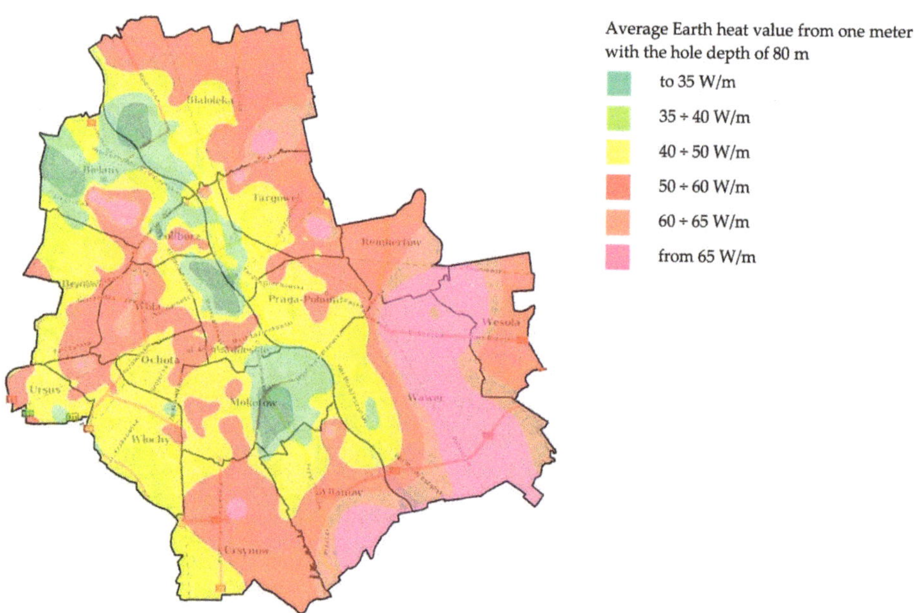


**Figure 9.** Low-temperature geothermal resources in Warsaw [8].

### 4.2. Solar Energy

Warsaw is located at 51 °N latitude in a moderate climate zone with influences of Atlantic and Continental climate, as is true of the whole country. Such a location causes the city (and Poland) to be affected by different atmospheric fronts that result in frequent heavy cloud formation. The averaged mean yearly temperature is equal to 7.9 °C and average annual global solar irradiation accounts for 1100 kWh/m$^2$ and solar hours are on average equal to 1600 [25]. The winters are relatively severe. The coldest months are January, February and December and solar irradiation is lowest in those months, which can be seen in Figure 10. In winter the ambient air temperature can even drop to −20 °C on extremely cold days. The average hourly ambient air temperature for averaged days of January and February (the coldest months in a year) varies during the daytime from −6 °C to −2 °C (January) or to 0 °C (February).

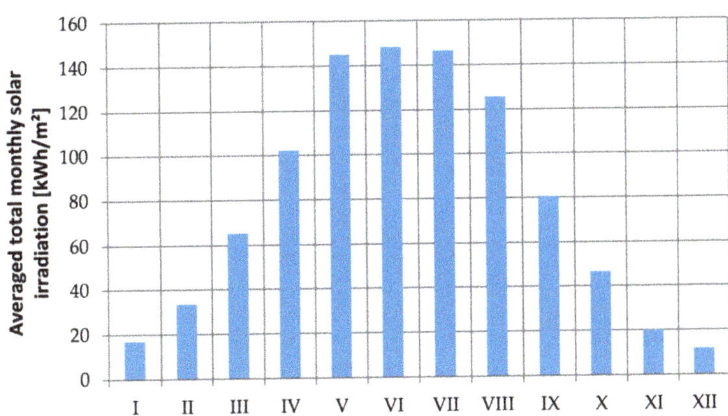

**Figure 10.** Distribution of averaged total monthly solar irradiation [kWh/m$^2$] on horizontal surface in Warsaw [25].

The highest solar irradiation occurs in June and in Warsaw the average monthly irradiation is nearly 150 kWh/m$^2$. The lowest solar irradiation occurs in December and the average monthly irradiation is equal to about 11 kWh/m$^2$. From October to April only about 20% of annual total radiation is available. The structure of solar radiation is characterized by a very high share of diffuse radiation. An average annual percentage of direct radiation amounts to only 46%. In summer, the share of direct radiation is higher and accounts for approximately 56%. However, from November to the end of February the percentage of diffuse radiation varies from 65% to 80%. This situation recommends solar systems for applying both direct and diffuse solar radiation; systems with concentrators are not recommended. With such a low share of direct solar radiation, the effect of concentrating solar beams is very small and can be seen only in summer. The small increase in efficiency is too expensive and the solar concentration technology (solar thermal power plants, including solar power tower plants) is not an economically efficient solution in Polish climate. The solar thermal heating systems consist of flat plate or vacuum tube solar collectors and water storage tanks. Anti-freezing mixture circulates in a solar collector loop. The annual share of the solar heating system supplying heat to DHW (domestic hot water) system can be at a level of 50%–65%; in the case of space heating the solar share is lower and accounts for 20%–30% (even when the surface area of solar collectors is at least two times bigger than in the case of a system only for DHW heating) [25].

It should be underlined that to determine the effective use of solar energy it is necessary to analyse the availability of solar energy. Estimation of distribution of solar radiation incident on any surface of solar energy receiver is very important. The surfaces of solar energy receivers, elements of active solar systems, i.e., solar collectors and PV modules, are tilted to the horizontal surface and can have different orientations (azimuth angle). To analyse solar energy availability, the appropriate solar radiation data are required. However, when we consider utilization of solar energy in cities, not only is access to reliable data important. The architecture of the buildings and the urban planning of the entire city is also very important. The arrangement of individual buildings and other construction structures in close neighbourhoods are also very important. The urban environment may limit the availability of solar radiation, which is related to the shading of the surface of solar receivers. In some locations, building walls and roofs can have very good exposure to incident solar radiation, which makes utilization of solar radiation very effective throughout the year. Figure 11 shows a solar map of the main campus of the Warsaw University of Technology. The figure shows the annual solar irradiation of the individual campus buildings, which enables efficient planning and dimensioning of the solar energy receiver locations. Consequently, it is possible to estimate the technical potential of using solar energy to generate heating and cooling energy and electricity in a given area or district of the city.

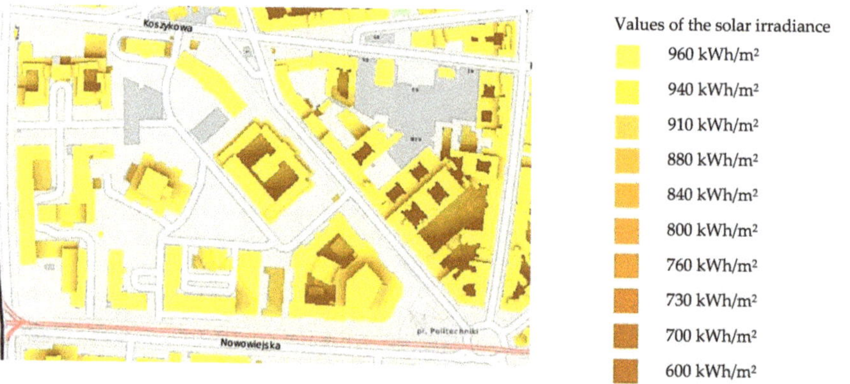


**Figure 11.** Solar map (Warsaw University of Technology, Main Campus as an example) [8].

It seems quite interesting to use solar radiation energy to generate electricity in some of the unshaded areas of facilities and transport facilities, which account for about 10% of the city's area. It should be noted that in the current legal and ownership situation, this energy can be mainly used by the landowner.

### 4.3. Biomass and Biogas

In the year 2019, a total of approximately 668,600 tons of municipal waste were produced, including 610,700 tons from households [26]. This means that a statistical resident of Warsaw generated 342.4 kg of municipal waste, including 271.1 kg of mixed waste. The share of mixed waste in the total mass of municipal waste collected was as high as 71.9%. Collection of biodegradable waste amounted to 27,098 tons and 15,020 tons of paper and cardboard. The calorific value of municipal waste ranges from 7–16 MJ/kg [27].

The production of electricity and heat from biogas from sewage is carried out in the "Czajka" and "Południe" sewage treatment plants [28]. Czajka's annual heat production is about 40 GWh with similar amounts of electricity. Annual heat production of the second plant is about 5 GWh. Both plants use the heat and energy for their own purposes.

Interestingly, the owner—Municipal Water and Sewerage Company in the capital city of Warsaw S.A.—plans to invest in photovoltaic installations in the near future, for a total production of 6.7 GWh per year [28].

Electricity and heat are also generated from municipal waste in the Municipal Waste Disposal Plant (owner: Municipal Cleaning Enterprise in the Capital City of Warsaw, location: Targówek district). The reconstruction of this plant is planned in the near future [10].

PGNiG Termika's plans also include a modern municipal waste incineration plant at the Siekierki heat and power plant [29].

What is important is that only the latest technologies will be applied in the new or reconstructed plants [10].

Every city generates a lot of waste. This waste should be recycled, both for the recovery of raw materials and for energy recovery. Each city may, or rather, should, strive to implement a circular economy, which is one of the most important and even basic elements of sustainable development and gives the base for PEDs creation.

### 4.4. Wind Energy

The possibilities of using wind energy to generate electricity in urban areas are quite limited due to the legal requirements in force, i.e., the location of the wind farm requires appropriate provisions in the local spatial development plans. In the part of the city under conservation protection, the installation of even micro wind farms is practically impossible. Availability of wind energy in the area of the Capital City of Warsaw's electricity production is relatively low.

One of the important parameters to urban meteorological investigations is an aerodynamic roughness length (RL, Figure 12 [30]). In the centre of the town, that value is equal to 1 m or in some urbanized areas is even equal to 1.5 m, while is most attractive in terms of wind energy utilization in parts where the value is much lower, e.g., in agricultural areas or wastelands along the Vistula River the factor is equal to 0.1 m. The minimum value of this indicator is above the river surface. In this area, the value is 0.0002 m.

**Figure 12.** Aerodynamic roughness length of the Warsaw's area [30].

In Warsaw, for the 10% of the windiest surface at a height of 10 m above the ground, the average density of the available power is 87 W/m$^2$ with an average air flow velocity of 4.00 m/s [30]. At a height of 50 m, these values are, respectively, 204 W/m$^2$ and 5.63 m/s. At a height of 100 m, these values are, respectively, 299 W/m$^2$ and 6.60 m/s. The best conditions are on the outskirts of the city, in the north and south, and along the Vistula, i.e., where the roughness length (RL) is not greater than 0.1 m, as can be seen on Figure 12.

The northern region (at the top of Figure 12) does not seem to be attractive for the use of wind energy due to the planned housing development.

The most favourable and sparsely inhabited area of about 17.0 km$^2$ (south, where the roughness length is not greater than 0.1 m, at the bottom of Figure 12) at a height of 10 m is characterized by an average wind speed of 4.4 m/s, and an average power density of 117 W/m$^2$. When at a height of 50 m the average wind speed is estimated as 5.7 m/s, and the average power density as 213 W/m$^2$. For the height of 100 m these values are the following: 6.7 m/s and 300 W/m$^2$, respectively. Based on the average power density and the assumed area of land, electricity production can be estimated: at a height of 10 m—63 GWh, at a height of 50 m—105 GWh, and at a height of 100 m—160 GWh.

### 4.5. Hydropower

Warsaw is located on the Central Masovian Lowland on the Vistula River, with no access to the sea. Although within the city limits the Vistula is 28 km long, the use of hydropower in Warsaw is not taken into account due to inadequate conditions, i.e., low slope of the land and high density of buildings [31]. The use of smaller rivers, streams or existing water channels with low and variable water flow is currently not considered.

It is worth adding that about 40 km to the north of Warsaw there is the Dębe Hydro Power Plant, which was established in the 1960s [32]. The installed capacity of this power plant is 20 MW, and the average annual energy production is about 91 GWh. This power plant directly supplies the Polish power system with energy.

## 5. Towards Positive Energy Districts in Warsaw and Risks to Achieving This Goal

As shown in the previous section, in Warsaw there is a theoretical potential for the use of renewable energies. It should be noted, however, that the use of renewable energies is a rational solution only if there is a reduction in energy demand for traditional methods, which in this case means reducing the energy demand of existing buildings or erecting new ones in accordance with the latest energy and environmental standards.

In Warsaw, buildings in terms of energy performance are very diverse. In many cases, it is possible to significantly improve their energy balance and reduce energy demand through refurbishment and thermo-modernization. Nevertheless, in the historic part of the city, as well as in densely built-up areas with too much energy demand for heating and ventilation, achieving the "positive energy" standard can be difficult even with the current state of technology.

In Polish climatic conditions, it is difficult to achieve energy self-sufficiency in urbanized areas based on only one type of renewable source because energy needs are too large in relation to the possibility of coverage from monovalent-source renewable systems.

The solar energy is easiest to acquire on non-shaded surfaces, e.g., roofs. South or south-west and south-east facades of buildings can also be used as long as they are not shaded.

Low-temperature geothermal energy cannot directly supply a high-temperature heating network. It may be necessary to create local or neighbourhood networks to distribute heat in a smaller area. Heat pumps will be required to raise the temperature. Later such local systems equipped with heat pumps can be connected, thus competing with the traditional district heating network.

The city of Warsaw "produces" large amounts of municipal waste. Some of them are not recyclable or biodegradable. However, they can be used in modern plants for the combined production of electricity and heat. Therefore, special attention can be put to developing modern landfill gas energy plants. Up until now none of these landfill plants operate in Warsaw or its surroundings.

Large areas of green spaces can also be a source of biomass. However, a traditional combustion procedure should not be applied. It is necessary to use highly effective gasification processes as a preliminary energy treatment of biomass residues.

Potential barriers for implementation of renewable energy technologies and achieving the status of a smart city with some positive energy districts should be identified and mentioned. These are mainly problems with ownership, financing new investments and refurbishment, and legal boundaries. Buildings may be private property, there may be business ownership issues, and buildings may belong to a cooperative, housing community or the city of Warsaw, or even a combination of the previously mentioned. This ownership structure determines the ways and possibilities of financing the retrofitting measures. It should be mentioned that land around buildings usually has a similar ownership structure. This, in turn, can result in problems with the placement of energy systems outside the building, and even applying extra thermal insulation to the building walls (the external size of the building increases). The use of renewable energy, rationalization of the energy efficiency of existing systems, proper energy, water and waste management are an important part of efforts to transform Warsaw into a smart city.

## 6. Conclusions

Sustainable development, carried out by taking into account the protection of the climate and the natural environment, is de facto a political, economic, technical and social problem even on the scale of the city.

Without the general awareness of the importance of these issues and, on the other hand, social pressure, activities in this area may not be accepted due to the high costs of implementing investments.

Moreover, the development of technology is still important in order to maximize the widely understood effectiveness of technical solutions. It is also important to prepare specialists who carry out comprehensive projects using renewable energy sources and take care of the proper operation and maintenance of existing systems.

Increased outlays for the implementation of the change of the heat and energy supply system require financial support, perhaps on preferential terms or from funds focused on renewable energy.

- The Capital City of Warsaw supports all activities aimed at reducing the consumption of conventional energy, increasing the use of renewable energy sources and reducing carbon dioxide emissions.
- Increasingly stringent requirements in the field of thermal and energy protection of buildings reduce both the demand for heating and ventilation and the required design power of heat sources.
- Lowering the required design heating load per area below 60 W/m² (indicative value) allows the use of low-temperature surface (e.g., floor) heating, which means that the heat source can be a heat pump using, for example, low-temperature geothermal heat available in Warsaw.
- Achieving a positive energy standard by parts of the city with buildings with unsatisfactory energy performance may be economically unprofitable with the current technical possibilities.
- Due to the diversified ownership structure of land, buildings and accompanying infrastructure, as well as in view of further city expansion plans, the easiest way to implement the idea of energy-positive districts is to modernize the relatively recently established housing estates and design new ones with the "obligatory" use of renewable energy sources.
- The use of renewable energy sources instead of conventional fuels means a significant reduction in carbon dioxide emissions and environmental protection.
- Changing the existing electricity and heat supply system for urbanized urban areas in Polish conditions requires comprehensive modernization of practically the entire city infrastructure.

Summarizing, it is very difficult to modernize large energy systems, which for many years have been the only ones supplying energy to various types of energy users in the city. However, it does not mean that nothing can be done. Warsaw, like many world cities, is expanding to new areas and this causes opportunity for development of new local energy systems based on renewables. Such systems can assure self-energy generation and consumption, giving energy security for new districts.

Moreover, new districts can share their energy with other neighbourhoods. As a consequence, the new districts can become Positive Energy Districts which will increasingly interfere with the central energy systems of the city centre. New PEDs will absorb central areas step by step, replacing ineffective central energy systems with new ones based on renewable energy, giving energy independence to the whole city.

It should be underlined that the way to achieve complete energy independence for the city is very complicated and rather long, which is what the authors have tried to present in this paper. If thermal energy (a central district heating system) and electricity (a central power grid) are available everywhere in the city, why does the energy sector, developers, communities and inhabitants have to look for new "fuels" and innovative energy conversion technologies, mainly those based on renewables? This is a question and idea which should be answered by the municipalities as often as possible. Of course, academic people and scientists and local "green" NGOs should help and support the urgent need of the development of local efficient energy generation systems based on locally available energy sources, such as renewables. They should foster the idea of sharing energy in a modern way, which gives the basis for creating more Positive Energy Districts in the city.

**Author Contributions:** Conceptualization, D.C. and H.J.; methodology, D.C. and H.J.; validation, D.C. and H.J.; formal analysis, D.C. and H.J.; investigation, D.C. and H.J.; resources, D.C. and H.J.; data curation, D.C. and H.J.; writing—original draft preparation, D.C. and H.J.; writing—review and editing, D.C. and H.J.; visualization, H.J.; supervision, D.C.; project administration, H.J.; funding acquisition, D.C. All authors have read and agreed to the published version of the manuscript.

**Funding:** The publishing procedure is funding by the EERA Smart Cities JP.

**Institutional Review Board Statement:** Not applicable.

**Informed Consent Statement:** Not applicable.

**Data Availability Statement:** The study did not report any data.

**Conflicts of Interest:** The authors declare no conflict of interest.

## References


1. Scott, R. *The History of the International Energy Agency—The first Twenty Years*; OECD Publishing: Paris, France, 1994.
2. Directive 2002/91/EC of the European Parliament and of the Council of 16 December 2002 on the energy performance of buildings. *Off. J. Eur. Commun.* Available online: http://data.europa.eu/eli/dir/2002/91/oj (accessed on 10 September 2021). 2003, L 001, 04/01/2003 65–71.
3. Regulation of the Minister of Infrastructure of 12 April 2002 on Technical Conditions to be Met by Buildings and Their Location. *J. Laws.* Available online: https://www.global-regulation.com/translation/poland/3353940/regulation-of-the-minister-of-infrastructure-of-12-april-2002-on-technical-conditions%252c-which-should-correspond-to-the-buildings-and-their-location.html (accessed on 10 September 2021). 2002 No.75, item. 690 (In Polish).
4. Ministry of Climate and Environment. Poland's Energy Policy until 2040. Available online: https://www.dziennikustaw.gov.pl/MP/2021/264 (accessed on 10 September 2021). Official Journal of the Republic of Poland Monitor Polski 2021, item 264. (In Polish)
5. Regulation of the Minister of Infrastructure and Development of 27 February 2015 on the methodology for determining the energy performance of a building or part of a building and energy performance certificates. *J. Laws.* Available online: https://www.iea.org/policies/12240-regulation-of-the-minister-of-infrastructure-and-development-of-february-27-2015-on-the-methodology-for-determining-the-energy-performance-of-a-building-or-part-of-a-building-and-energy-performance-certificates (accessed on 10 September 2021). 2015, No. 2019, item 1065, consolidated text as amended. (In Polish)
6. The City of Warsaw. *Infrastructure Office: Action Plan for Sustainable Energy Consumption for Warsaw by 2020*; Infrastructure Department of The City of Warsaw: Warszawa, Poland, 2013.
7. Council of the Capital City of Warsaw. *Resolution No. LIX/1534/2017of the Warsaw City Council of 14 December 2017*; Housing Policy—Housing2030; Warsaw City Council: Warsaw, Poland, 2017.
8. City of Warsaw. A Map of Renewable Energy Sources (RES)—A Map of the Sun Exposure, Low Temperature Geothermal Energy (Heat Pumps) and Range of the Heat within the Network. Available online: http://mapa.um.warszawa.pl/mapaApp1/mapa?service=mapa_oze&L=en (accessed on 19 June 2021).
9. Council of the Capital City of Warsaw. Resolutions of the Council of the Capital City of Warsaw No. XXXV/1074/2020. In *Assumptions for the Heat, Electricity and Gas Fuel Supply Plan for the Capital City of Warsaw*; Warsaw City Council: Warsaw, Poland, 2020. (In Polish)
10. Public Information Bulletin of the Capital City of Warsaw. Available online: https://bip.warszawa.pl/English/default.htm (accessed on 30 June 2021).
11. Council of the Capital City of Warsaw. Strategy for Adaptation to Climate Change for the City of Warsaw up to 2030 with a Perspective to 2050. In *City Adaptation Plan*; Project LIFE_ADAPTCITY_PL, Institute for Sustainable Development: Warsaw, Poland, 2019; Available online: http://adaptcity.pl/wp-content/uploads/2019/11/strategia_2030-ENG.png (accessed on 10 September 2021).
12. Low-Emission Economy Plan for the Capital City of WARSAW. Available online: http://infrastruktura.um.warszawa.pl/sites/infrastruktura.um.warszawa.pl/files/dokumenty/plan_gospodarki_niskoemisyjnej_dla_m.st_._warszawy.png (accessed on 30 June 2021).
13. History of Warsaw. Available online: https://en.wikipedia.org/wiki/History_of_Warsaw (accessed on 20 June 2021).

14. Statistics Poland, Area and Population in the Territorial Profile in 2020. Available online: https://stat.gov.pl/en/topics/population/population/area-and-population-in-the-territorial-profile-in-2020,4,14.html (accessed on 19 June 2021).
15. Administrative Division of Warsaw into Districts. Available online: https://warszawa.wikia.org/wiki/Dzielnice?file=Warzawa_podzial_2002.png (accessed on 20 June 2021).
16. Statistical Office in Warszawa, Statistical Review of Warsaw—1st Quarter 2021. Available online: https://warszawa.stat.gov.pl/en/communiqus-and-announcements/communiqus-and-announcements/other/statistical-review-of-warsaw-1st-quarter-2021,1,30.html (accessed on 21 June 2021).
17. Mazovian Regional Research Center. Statistical Office in Warsaw. Ranking of Warsaw Districts in Terms of the Attractiveness of Living Conditions. Available online: https://warszawa.stat.gov.pl/files/gfx/warszawa/en/defaultaktualnosci/810/2/1/1/ranking_dzielnic_warszawy_pod_wzgledem_atrakcyjnosci_warunkow_zycia_ang.png (accessed on 12 July 2021).
18. Council of the Capital City of Warsaw. Study of the conditions and directions of spatial development in the capital city of Warsaw. In *Annex to Resolution No. LXII/1667/2018 of the Council of the Capital City of Warsaw from 1 March 2018*; Council of the Capital City of Warsaw: Warsaw, Poland, 2018. (In Polish)
19. Ministry of Infrastructure and Development: National Plan to Increase the Number of Buildings with Low Energy Consumption. Monitor Polski 2015, item 614; Official Journal of the Republic of Poland: Warsaw, Poland, 2015. (In Polish)
20. National Energy Conservation Agency S.A. (NAPE). *Polish Building Typology—TABULA*; Scientific Report; Narodowa Agencja Poszanowania Energii SA: Warsaw, Poland, 2012.
21. Żerań CHP Plant. Available online: https://termika.pgnig.pl/elektrocieplownia-zeran (accessed on 12 July 2021).
22. Chwieduk, D.; Bujalski, W.; Chwieduk, B. Possibilities of Transition from Centralized Energy Systems to Distributed Energy Sources in Large Polish Cities. *Energies* **2020**, *13*, 6007. [CrossRef]
23. Socha, M.; Sadurski, A.; Skrzypczyk, L. Possibilities of geothermal energy utilisation in the Warsaw urban area on the background of cost analysis. *Prz. Geol.* **2016**, *64*, 7.
24. Szewczyk, J. Geophysical and hydrogeological aspects of utilization of thermal energy in Poland. *Prz. Geol.* **2010**, *58*, 7. (In Polish)
25. Chwieduk, D. Solar Energy Use for Thermal Application in Poland. *Polish J. Environ. Stud.* **2010**, *19*, 473–478.
26. Statistics Poland. Local Data Bank. Available online: https://bdl.stat.gov.pl/BDL/start (accessed on 10 August 2021). (In Polish)
27. Wielgosiński, G. *Thermal Waste Treatment*; Nowa Energia Publishing House: Racibórz, Poland, 2020. (In Polish)
28. Zero Waste—How to Minimize Waste Production. Available online: http://www.um.warszawa.pl/aktualnosci/zero-waste-jak-maksymalne-zmniejszy-produkcj-odpad-w?page=0,1 (accessed on 30 June 2021).
29. PGNiG Termika's Strategy for 2015–2022. Available online: https://termika.pgnig.pl/sites/default/files/2016-07/Strategia%20PGNiG%20TERMIKA%20na%20lata%202015%20-%202022.png (accessed on 30 June 2021).
30. Global Wind Atlas. Available online: https://globalwindatlas.info/ (accessed on 10 August 2021).
31. Warsaw. Available online: https://en.wikipedia.org/wiki/Warsaw (accessed on 10 August 2021).
32. Dębe Hydroelectric Power Plant. Available online: https://pgeeo.pl/Nasze-obiekty/Elektrownie-wodne/Debe (accessed on 10 August 2021).

*Article*

# Combining Sufficiency, Efficiency and Flexibility to Achieve Positive Energy Districts Targets

Silvia Erba * and Lorenzo Pagliano

eERG, End-Use Efficiency Research Group, Politecnico di Milano, 20133 Milano, Italy; lorenzo.pagliano@polimi.it
* Correspondence: silvia.erba@polimi.it

**Abstract:** Energy efficiency, generation from renewable sources and more recently energy flexibility are key elements of present sustainability policies. However, we are beginning to see a recognition of the need to couple technological solutions with lifestyle and behavioral changes, sometimes labeled under the term "sufficiency". Appropriate policies and design principles are necessary to enable sufficiency options, which in turn reveal that there is a bidirectional influence between the building and the district/city level. In this context, the authors discuss how city and building re-design should be implemented combining energy efficiency, flexibility, production from renewables and sufficiency options for achieving a positive energy balance at the district level even within the constraints of dense cities. Based on a review of recent advances, the paper provides a matrix of interactions between building and district design for use by building designers and city planners. It also compares possible scenarios implementing different strategies at the building and urban level in a case study, in order to evaluate the effect of the proposed integrated approach on the energy balance at yearly and seasonal time scales and on land take.

**Keywords:** energy sufficiency; deep energy retrofit; energy flexibility; energy efficiency; building thermal mass; positive energy district; yearly energy balance; seasonal energy balance

**Citation:** Erba, S.; Pagliano, L. Combining Sufficiency, Efficiency and Flexibility to Achieve Positive Energy Districts Targets. *Energies* **2021**, *14*, 4697. https://doi.org/10.3390/en14154697

Academic Editors: Paola Clerici Maestosi, Chi-Ming Lai and Patrick E Phelan

Received: 1 May 2021
Accepted: 21 July 2021
Published: 3 August 2021

**Publisher's Note:** MDPI stays neutral with regard to jurisdictional claims in published maps and institutional affiliations.

## 1. Introduction

The challenge of sustainable development of urban areas is of key importance for the European Union, which has defined an ambitious strategy and implementation plans to make cities inclusive, safe, resilient and sustainable, in accordance with the 2030 Agenda for Sustainable Development of the United Nations [1]. Ongoing and foreseen accelerated urbanization in many areas of the world interacts with other challenges, including overpopulation, climate change, environmental quality and access to energy [2]. Urban and regional planning is called to reassess how to sustainably supply the population with the needed services at an affordable cost. Reflecting this, urban actors and scholars have created a number of city labels, such as "sustainable city", "smart city", "green city" and "resilient city", to represent cities' responses to various challenges of urban transformation. Among them, the 'smart city' has prevailed as the most researched concept in the recent period [3], even if it is sometimes presented as being focused only on the application of information and communication technologies (ICTs) and leaving in the background a number of issues related to building a physics and space and social organization. Most recently a "15 min City" concept has been presented that proposes fundamental changes in urban planning aimed at redesigning neighborhoods so that individuals can reach the school, workplace, groceries, sport and recreational sites, etc., within a 15 min travel distance, either by bike or on foot [4,5]. Each neighborhood should fulfil six social functions: living, working, supplying, caring, learning and enjoying. The concept may have relevant implications on energy and material use not only in the area of mobility, but also elsewhere, as we will discuss in the paper.

The EU has been investing in sustainable urban development research for over twenty years [6] and has recently announced the mission [7] to guide the transformation of 100 European cities to climate neutrality by 2030, supporting the cities through different financial means, e.g., the new framework program Horizon Europe.

In particular, the area of "Smart City & Community" has been defined as strategic and a priority since the previous European Horizon 2020 Program, which has funded numerous projects to foster European Smart cities and communities, e.g., [8,9]. Over time, however, it has been realized that financing large smart city projects at the urban level was complex and with a huge demand for resources and investments. As an intermediate step, to be achieved in a shorter time frame, a focus has been developed towards smaller urban areas, such as districts.

In June 2018, the European Strategic Energy Technology Plan (SET Plan), based also on previous planning, has proposed an implementation plan on "Smart Cities and Communities" dedicated to develop 100 smart positive energy districts (PEDs) in Europe by 2025 (Action 3.2), characterized by improved sustainability, livability and going beyond carbon neutrality. The Program on Positive Energy Districts and Neighborhoods (PED Program), led by the intergovernmental Joint Programming Initiative (JPI) Urban Europe, has been established to support this ambitious action, and has realized a review of early attempts to PEDs in Europe [10]. The district approach is mentioned also within Article 19 of the revised Energy Performance of Buildings Directive (2018/844) [11], which, also for the first time, does not limit to energy use of buildings but underlines the link between buildings, mobility and urban planning (recital 28 and article 8). Finally, the Horizon 2020 work program [12] stresses the importance of deploying positive energy blocks and districts by 2050 in Europe to achieve the needed energy transition in cities, in addition to foster the integration between energy systems and improve the buildings' energy performance significantly beyond the levels of current EU codes.

Within the dedicated calls of the H2020 program, a definition for positive energy blocks and districts is given: "Positive Energy Blocks/Districts consist of several buildings … that actively manage their energy consumption and the energy flow between them and the wider energy system. Positive Energy Blocks/Districts have an **annual** positive energy balance".

The PED Program has elaborated a framework definition, which also uses the concept of **yearly** energy balance and extends from the urban to **regional level** the boundary of the system where a positive value of the balance should be achieved: "Positive Energy Districts are energy-efficient and energy-flexible urban areas, which produce net zero greenhouse gas emissions and actively manage an **annual local or regional** surplus production of renewable energy." [13]. However, according to Lindholm et al. [14] the concept of a positive energy district is still in an early conceptual phase and research to exploit its value shall be taken into account. PEDs may offer interesting replicability and scalability potentials [15], thus it is crucial to identify clearly objectives, strengths and opportunities, via explicitly and univocally defined indicators and with an explicit and structured calculation methodology [16]. A strong critique, backed by detailed optimization calculations, of the **yearly net zero** energy metric is presented in [17]. The authors apply optimization for example to a district made up of well-insulated apartments, heated by a heat pump and endowed with PV panels and an electric battery. By optimizing with the objective of maximizing the net generation over a year they find that the battery would not be utilized, since "battery losses will result in net increase in electricity consumption compared with the no storage case" hence damaging the yearly energy positivity goal. A conflict arises with the objective of maximizing the use of renewable energy at the time when it is available, that is with the objective of being "flexible" about the time when the district uses energy, either for direct production of services or for storage and delayed use.

Comparing the above PED definitions, it can be observed that in the definition by the PED Program a threefold objective is highlighted: energy-flexibility (though not quantitatively characterized), the target of positive energy balance and that of net zero greenhouse gas emission, which recalls the concept of the zero emission neighborhood (ZEN) [18,19]. Regarding the target of net zero greenhouse gas emissions, it is worth noting that of the

29 districts in Europe that declared a PED ambition in the booklet under development by JPI Urban Europe [10], 19 cases indicate at least one target between carbon neutrality, zero emission or climate neutrality. According to the definitions reported in the IPCC Special Report: Global Warming of 1.5 °C [20], a carbon neutral goal refers to carbon dioxide only, whereas a 'net-zero' target includes all greenhouse gases, and a 'climate-neutral' goal extends to all causes of radiative forcing.

The framework definition distinguishes three main functions related to PEDs: energy efficiency, energy flexibility and energy production.

Currently, energy efficiency and, more recently, energy flexibility drive the policy practices to achieve high sustainability goals, e.g., a clever utilization of thermal mass may allow one to manage the building as a thermal battery over a time frame of a few days rather than hours, if the building fabric is highly insulated and high efficiency heat recovery on ventilation is applied (as analyzed in Section 3 of this paper). Therefore, in the case of existing buildings, the path of deep renovation focused first at improving the building fabric can be a prerequisite enabler. This flexibility (in new or retrofitted building fabrics) allows dealing with the challenges linked to the intermittent nature of many renewable energy sources and their exploitation at the level of a cluster of buildings. Given the need of a means of storage from the daily to the interseasonal scale, a strong reduction of *energy needs for heating and cooling* via efficiency techniques and physical and regulatory frameworks that enable low-energy life-styles (i.e., sufficiency, discussed in detail in Section 4) might prove decisive. This would reduce the size of the required storage and the connected embedded energy and energy losses.

At the same time, a strong reduction of *energy needs* and hence of the physical infrastructures required to serve those needs, appears as a fundamental step for achieving the European and international goals related to halting land consumption. The United Nations Sustainable Development Goal (SDG) indicator 11.3.1: "Ratio of land consumption rate to population growth rate" postulates that when this ratio is high, such a "growth turns out to violate every premise of sustainability that an urban area could be judged by". In Europe, where population is projected to remain stable or even slightly declining throughout this century [21], the EU institutions have taken a commitment to be "a frontrunner in implementing [ . . . ] the SDGs" and to aim at "no net land take by 2050" [22,23]. On 29 April 2021 the European Parliament approved with a majority of 605/660 a resolution asking the EU Commission to draft a new directive for the protection of soil with the objectives of "no land degradation" by 2030 and "no net land take" by 2050 at the latest.

Essential to reach the transformation of cities is the involvement of citizens and stakeholders because of their central role in interacting with the buildings and the district's infrastructures. In this regard, we are beginning to see a recognition (e.g., in some of the H2020 calls, in the IEA outlook 2020 [24], in chapter 5 of the upcoming IPCC report) of the need to couple technological solutions with lifestyle and behavioral changes, sometimes labeled under the term "sufficiency". However, sufficiency is not simply an issue of choices of each individual: sociological and psychological research indicates the need for enabling infrastructures and social frameworks [25,26]. Hence, appropriate policies and design principles are necessary to enable sufficiency options, which in turn reveal that there is a bidirectional influence between the building and the district/city level.

In this context, the authors discuss how city and building redesign should be implemented combining efficiency, flexibility and renewables production with the emerging new dimension of sufficiency options for achieving a positive energy balance at the district level even within the constraints of dense cities (Figure 1). The paper focuses on the buildings related aspect of the energy district while transports, public spaces and mobility are discussed in terms of their interaction with building infrastructure.

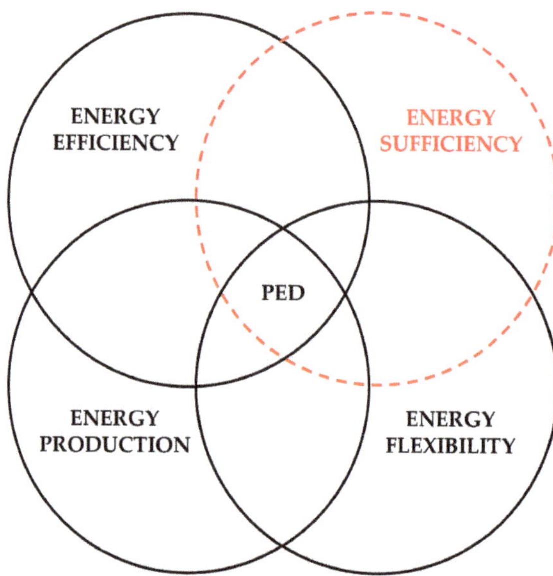

**Figure 1.** Schematization of the functions of a PED.

## 2. Energy Efficiency

One of the key elements frequently present in the current PED concepts is that the district generates more renewable energy than it consumes on a **yearly** basis. This is achieved by integrating renewable energy systems and energy storage and improving the energy efficiency of the district [14].

An energy efficient district will take into account the energy uses of different sectors and building typologies and it will be constituted by mixed use and relatively dense settlements in order to minimize transportation needs.

Further, it will be characterized by buildings with low *energy needs for heating, cooling and hot water*, low *energy use* for lighting and ventilation, energy efficient building service systems, smart energy management and technologies and energy efficient street lighting.

The directive 2012/27/EU [27] requires all EU countries to use energy more efficiently at all stages of the energy chain, including energy generation, transmission, distribution and end-use consumption. In the 2018 recast of the directive [11], the Commission proposed an ambitious energy efficiency target by 2030, regarding final energy use (whichever the source, renewable or non-renewable) and primary energy, and more stringent revisions are ongoing.

In addition, specifically for the building sector, the EU has established the Energy Performance of Building Directive 2010/31/EU [28], amended in 2018 [11], which sets out policies and supportive measures to improve the buildings energy performance and upgrade the existing building stock. The directive requires all new buildings (and major retrofits) from 2021 (public buildings from 2019) to be nearly-zero energy buildings (nZEBs), defined as buildings that:

- Have a very high energy performance;
- Cover, to a very significant extent, the nearly zero or very low amount of energy required by energy from renewable sources (including energy from renewable sources produced on-site or nearby).

The detailed choice of indicators, rather than only of the numerical values to be achieved, has been left to Member States. The resulting national implementations of the nZEB concept are hence considerably different in terms of the selected indicators, which makes it difficult to compare nZEBs in different countries [29–34].

EU has also promoted research dedicated to find out and clarify the best strategies and technological solutions to make the nZEBs affordable, in order to allow a rapid market uptake and thus helping to achieve the EU's energy and environmental goals [32,35].

According to ISO 52000-1 [36], the indicators to assess and design a nZEB should be three and should be considered in the following order: (1) *energy needs for heating and cooling*, to reflect the performance of the *building fabric*, quantifying and promoting the reduction of energy losses through the envelope and ventilation; (2) *total primary energy*, to reflect the performance of the technical building systems; (3) *non-renewable primary energy* for quantifying and promoting the reduction of the non-renewable fraction within *total primary energy* use. Within the AZEB project, the authors have developed a series of simplified graphical illustrations (Figure 2) and a video to show in a clear and concise way the above concepts and nomenclature [35]. The indicators *energy needs* and *total primary energy* do respond to the energy efficiency first (EEF) principle, which is one of the key principles of the Energy Union, intended to ensure secure, sustainable, competitive and affordable energy supply in the EU. The parameter *non-renewable primary energy* responds to the objective of "increasing the share of renewables". Reducing *energy needs* will not reduce the necessity of (and the market for) renewables and controls. On the contrary, it constitutes an indispensable prerequisite for these to be deployed with effective and acceptable results from the social and environmental point of view, including the EU objective of zero "land take", and therefore for their rapid penetration.

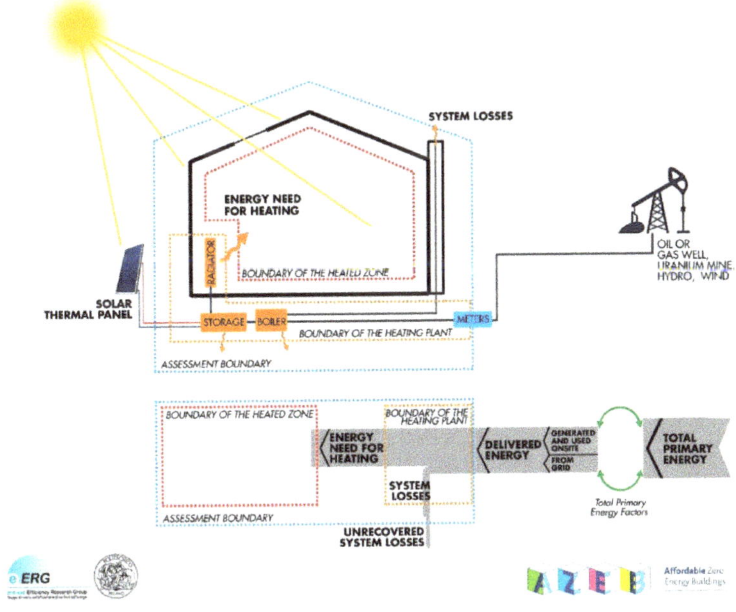


**Figure 2.** Graphical representation of energy levels developed by the authors (S. Erba and L. Pagliano): case where the energy service considered is space heating, delivered by a boiler and on-site solar thermal panels.

To move beyond the concept of nZEB to that of PED, it should be followed a similar rational sequence of steps, starting by reducing the overall *energy use* of the district, then covering this use fully with renewables, by making it flexible in order to accommodate short term fluctuations in RES supply, creating an interseasonal storage infrastructure, and providing a set of RES coherent with the territory. On this line, the technical report by the European Commission Joint Research Centre [37] suggests to extend the application of the minimum energy performance requirements by the EPBD from the building's to the cluster of buildings' level, keeping the priority on the EEF principle.

Different examples of energy efficient districts are spreading in Europe, characterized by the presence of high performance buildings, which act as a starting point to reduce the overall energy use of the neighborhood, e.g., in the Carquefou district (Nantes), France-Fleuriaye West, all of the new residential buildings are certified Passivhaus [38] and due to the correct orientation of the housing units, the careful design of the envelope, the proper management of solar heat gains and the exploitation of night ventilation, they are able to minimize the energy needs for heating and cooling guaranteeing thermal comfort throughout the year. The choice of energy efficient equipment allows one to further reduce the electric energy use. In Bærum, Norway-Fornebu [39], all the buildings will be nZEB, with the minimum passive house standard and energy class A. In addition, all buildings should be certified according to BREEAM-NOR Excellent. In Bodø, Norway-NyBy Buildings are planned to be built according to the ZEB standard [40].

In a PED not only new constructions but also existing buildings need to be addressed. Presently, roughly 75% of the EU building stock is energy inefficient and the poorly insulated buildings are not suitable for effectively helping the logic of positive energy districts. "Deep renovations" are a necessary condition to allow the building to modulate demand and provide flexibility to the network, as we discuss in Section 3. A unique definition of "deep renovation" or "deep energy renovation" is still not available; however, the literature offers studies and benchmarks that allow one to identify the main targets of this strategy [41–43]. A report by the European Parliament (30 July 2012), states that "deep renovation means a refurbishment that reduces both the delivered and the final energy consumption of a building by at least 80% compared with the pre-renovation levels" [44].

To reach and verify these goals, the process should start with the assessment of the existing building performance, proceeding with the design (preferably the codesign in multiproperty buildings), the construction, the commissioning, the management and operation of the building. Recent deep retrofits have proven the possibility to reduce the *energy need for heating* to 15–20 kWh/(m$^2$·y) even with a limited budget as often is the case in social housing [8]. However, concern has been expressed about the effectiveness of energy efficiency upgrades in bringing about the level of energy savings foreseen at the design stage [45]. The energy performance gap and prebound and rebound effects are often reported when dealing with energy efficiency upgrades and there is relatively ample literature on the subject [46]. To measure and verify the actual building performance, i.e., the energy performance and the indoor environmental quality (IEQ), measurement and verification (M&V) protocols should be implemented after the design phase of a new construction or a building retrofit. Since 1996, when EVO published the so called "International Performance Measurement and Verification Protocol" (IPMVP) [47], different guidelines and procedures have been developed to verify that a building performs according to design expectations, suggest energy saving measures to be adopted to further improve the building performance or to facilitate operation and maintenance. Specific work has been developed to define M&V plans for ZEBs [48] and more recently addressing zero energy settlements [49]. One of the increasingly discussed solutions to mitigate the performance gap is to enhance the awareness of the consumer and encourage a more efficient and sustainable energy-saving behavior at the building [50] and district level. It should be noticed that some measurement campaigns show good agreement between the predicted and actual performance in the case of Passivhaus buildings, which might be a consequence of a particularly accurate design and execution under this voluntary label. A recent report by the Universite de Geneve and SUPSI [51] finds that for the buildings labeled Minergie A and Minergie P, "the analysis yields a negative Energy Performance Gap (EPG) of −14% (i.e., the median building consumes slightly less than its standard), which provides further support for the initial hypothesis that the most efficient buildings are more robust to the EPG".

Currently there is no formalized standard to calculate an energy balance applied at the district scale but the literature offers the first attempts to define methodologies for the development of positive energy district targets and optimization techniques [16,37]. According to Shnapp et al. [37], the EPBD's cost-optimality calculation methodology, set up to calculate the minimum performance of individual buildings in MS regulation, might be adapted to establish minimum district performance requirements for zero or positive energy district solutions. We notice here that presently the cost-optimal methodology does not include external costs (including damage due to local pollution and climate change), which are large and increasing, and hence its adaptation to a wider context should in our view include explicitly those "negative externalities" in its calculation methodology [52,53].

Gabaldón et al. [16] have proposed a methodology to support cities at the design stage of a district evaluating its annual energy balance. It follows the approach of the ISO 52000 standard and it is made of eight steps: after the definition of the PED boundaries, the *energy needs* are calculated and followed by a series of iterative steps that finish calculating the equivalent primary energy, the primary energy balance and an associated Sankey diagram. The primary energy balance is defined in the paper as the difference between the *non-renewable primary energy* delivered to the district and the *renewable primary energy* that is exported outside the PED's boundaries. The authors also warn that "Because it is possible to achieve a Positive Energy District through an unrealistic system (as huge on-site generation, to give an example), a standardized calculation is needed similar to the nZEB rating, where lowering the *energy needs* is a first requisite to certifying the building (and by extension a district)".

Both at the district and building level a general critique holds with respect of an energy balance performed over a year when compensation is allowed between different time steps: it allows one to compensate for continued fossil use in some moments (e.g., winter) with overproduction of RES at other times (e.g., summer) This obviously does not lead to zero emissions and requires potentially large interseasonal storage locally or somewhere in the network, with associated land occupation, energy losses and use of materials, embedded energy and related costs. Additionally, the fact of choosing a long time interval for the energy balance creates a conflict with other objectives, e.g., the self-consumption of renewables as already mentioned [17]. As for buildings, some MS such as Italy and Spain have opted for running the balance with only partial compensations, meaning that compensation happens within months but not between different months. Additionally, in fact lower time periods for compensation might be chosen in case more detailed calculation methods [54] would be adopted for the energy certification (e.g., when hourly calculation methods would be adopted) rather than the most current monthly calculation method. A similar method to limit the negative effects of the simplistic yearly balance should be adopted also in performing the energy balance of a district. We offer an example of the issues in the case study presented in Section 5.

### 3. Energy Flexibility

Demand for energy flexibility is increasing due to the growing penetration of variable renewable energy sources into the energy system. Among non-dispatchable sources, solar power and wind power contribute noticeably to the electrical grid and a generation above a 30% share in annual electricity consumption strongly increases flexibility requirements [55]. This creates a challenge for the energy use in buildings to become in the future flexible, capable to adapt to the needs of the network and renewable production, maintaining standards related to comfort and low operating costs.

Different ways of obtaining energy flexibility can be listed such as the building thermal mass, which can be utilized to store energy, active storage systems such as domestic hot water (DHW) storage and buffer tanks, chemical batteries, connection to more energy networks, etc. [56].

The revised EPBD [11] highlights the importance of buildings' energy flexibility in relation to the development of a smart readiness indicator, which will assess the technological readiness of buildings to adapt in response to the occupant's needs, to be flexible according to the needs of the grid and to facilitate maintenance and efficient operation.

It requires that one of the key functionalities of smart readiness in buildings is "the flexibility of a building's overall electricity demand, including its ability to enable participation in active and passive and implicit and explicit demand-response, in relation to the grid, for example through flexibility and load shifting capacities.".

Individual buildings or clusters of buildings can provide flexibility. Annex 67 [56] defines the energy flexibility of a building as "the ability to manage its demand and generation according to local climate conditions, user needs, and energy network requirements. It will thus allow for demand side management/load control and thereby demand response based on the requirements of the surrounding energy networks". Compared to a single building, the flexibility and use of new technologies can in principle be increased by focusing at the district level [57,58] since this might ensure a larger accommodation of RES supply systems and easily flatten the load profiles due to the presence of different occupancy patterns and a potential variety of RES sources and heat sinks (e.g., soil, water bodies, night sky and night ventilation in summer). Besides, a larger scale might offer economies of scale in the case of retrofit actions aggregating groups of buildings and options for energy production and storage, which may not be economically or practically suitable in the case of a single building.

In the literature, a consolidated definition of the energy flexible district is missing; however, different concepts are available to describe the synergy of energy efficient buildings and renewable energy utilization at an aggregated level. In particular, Vigna et al. [59] focused their review on the characterization of the concept of energy flexible building cluster, defining it as a group of buildings interconnected to the same energy infrastructure, which should be able to manage their generation, storage and consumption in response to forcing factors with the aim to exploit as much as possible RES while reducing $CO_2$ emissions. As noted, those objectives are not coincident with the objective of net (or positive) zero energy over a year.

Among the different systems that can compete to determine energy flexibility we highlight in this paper the potential for energy storage of the buildings' envelope and structural elements, which allows it to accumulate energy when the demand is low in order to reduce it during peak periods. The aim of this analysis is to show the energy demand management ability of different types of buildings to allow preliminary evaluations of the potential within a district, which can include a mix of highly insulated buildings with large time constants and poorly insulated buildings with smaller time constants (as it might happen in historical centers where some buildings present limitations for external insulation). Different studies have shown the potential of structural thermal mass in achieving flexibility. The majority of the studies uses dynamic modeling to identify this potential [60–70]. Fewer studies have been implemented through the realization of experimental campaigns [71–74]. We provide in this paper (Section 5) the findings from a case study of a multifamily residential building located in Milan, Italy, which has been assessed via modeling and verified through experimental tests. We chose to focus the attention on the potential linked to the thermal mass since it can be considered as a "passive" thermal storage, whose cost is already paid off by energy savings, requiring just one extra investment, i.e., appropriate controllers of the heating system.

The principle underpinning the exploitation of the building thermal mass, e.g., in winter is to increase the building set point indoor temperature to accumulate heat when energy/electricity locally generated by renewables is available and to decrease it when the power production is too low. However, at the same time, it is necessary to guarantee adequate comfort conditions for the users, which can be controlled by keeping the operative temperature within the limits of the occupants' thermal comfort range as derived from comfort models and made explicit, e.g., in comfort standards (EN 16798 [75] and ASHRAE 55 [76]).

Le Dréau and Heiselberg [63] evaluated the use of thermal mass to modulate the energy use of the heating system, comparing poorly versus highly insulated buildings through energy modeling. They showed that poorly insulated buildings can modulate a large amount of heat only for short periods of time (2–5 h). Contrarily, well-insulated buildings are able to modulate a smaller amount of heat but they can maintain acceptable indoor comfort conditions after a complete switch-off of the heating systems for more than one day. Similarly in [67] they applied an energy flexibility index to show that poorly insulated houses are less energy flexible than well insulated ones. A proper level of thermal insulation allows one to reduce the *energy needs for heating* and to recover a greater share of the accumulated thermal energy. Their analyses show that the insulation level in a building has the key role in determining its energy flexibility compared to the building's total thermal inertia, the type of heat emitter or the kind of additional indoor thermal mass. Additionally, Foteinaki et al. [61] have highlighted similar conclusions quantifying, e.g., in more than two days, the period of time during which a low-energy single-family house is able to maintain the temperature above 20 °C, after the heat supply is interrupted.

Currently, a significant part of the existing building stock is still characterized by buildings with low energy performance, which in winter require energy generally in the same morning time slot after the night attenuation. For this reason, it is important to couple this strategy with the deep energy retrofit of the buildings.

## 4. Energy Sufficiency

The concept of "sufficiency" was introduced by a number of researchers both from the energy and the sociology fields, following the energy crises of the 1970s and 1980s [77,78]; in the 1990s it was brought into the sustainability debate by W. Sachs: "A society in balance with nature can in fact only be approximated through a twin-track approach: through both intelligent rationalization of means and prudent moderation of ends" [79] and "While efficiency is about doing things right, sufficiency is about doing the right things" [80]. Since then the concept has been the subject of a rather large body of academic research [81], sometimes under different terminology frameworks. A recent review of concepts and terminology is offered in [82], which summarizes: " . . . Samadi et al. [83] make a distinction between efficiency, consistency and sufficiency defined as follows: "efficiency is an option in which the input-output relation is improved . . . consistency aims at fundamental changes in production and consumption by substituting non-renewable resources with renewable resources . . . [and] sufficiency is linked to the level of demand for goods and services". This distinction can be compared with the distinctions of the avoid-shift-improve (ASI) framework [84]: improve matches with efficiency and technological substitution (i.e., consistency), while shift and avoid correspond to lifestyle change (i.e., sufficiency)" (see Table 1).

**Table 1.** Schematic comparison of different types of behavior changes, adapted by authors from [82].

| Integrated Assessment Models (IAM) Distinction | Efficiency | (Technological) Substitution | Lifestyle Change | |
|---|---|---|---|---|
| | **EFFICIENCY** | **CONSISTENCY** | **SUFFICIENCY** | |
| Transport | Fuel-efficient vehicles | Vehicles powered from RES | Public transport | Teleconferencing, walking, cycling |
| Residential | Energy-efficient appliances (high level in energy labeling) | On-site generation by RES | Thermostat adjustment | Smaller apartments, reduced number and size of appliances |
| Consumer goods and services | Efficient supply chain | Purchase sustainable goods | Sustainable use of goods | Sharing goods |
| | Improve | | Shift | Avoid |

Recently, the sufficiency concept has been incorporated as a key element into the Energy Plan 2020–2030 of the State of Genève, under the French name "sobrietè" [85], it has been included into the French Law for Energy Transition (2015) on equal footing with energy efficiency and has appeared as an element of future energy scenarios in the analysis of international bodies officially appointed to deal with energy (IEA) and climate (IPCC).

The latest World Energy Outlook by the International Energy Agency introduces explicitly sufficiency actions, described there with the term "behaviour changes", and their effect on energy use between 2020 and 2030 [24]. IEA included in its list various sufficiency actions, e.g., changing the thermostat settings for summer and winter, line-drying clothes, walking and cycling, working from home, car-sharing, etc. The next Advancement Report (AR6) by the Intergovernmental Panel on Climate Change (IPCC) is expected to cover demand-side solutions in a new chapter (Chapter 5 of the WGIII: Demand, services, and social aspects of mitigation) where "demand refers to end-use demand for services, such as nutrition, mobility, thermal comfort and lighting. It emphasizes services rather than consumption as essential dimension to guarantee constituents of wellbeing" [84,86]. Concrete examples of communities embracing the concept of physical limits, rather than simply of "doing more with less" via technical efficiency, are, e.g., the cities of Amsterdam and Brussels, which have adopted the "doughnut" concept proposed by ecological economist Kate Raworth [87] in which the outer ring of the diagram represents Earth's environmental ceiling, a place where the collective use of resources has an adverse impact on the planet. Lower energy use, though, should not be confused with lower welfare levels, nor with a concept of restriction or deprivation, either voluntary or imposed [88]. A wealth of research shows that the growth in the use of energy and materials has in many countries reached levels where this use becomes dysfunctional and detrimental to general and individual welfare [89,90], due to its impacts on, e.g., in the case of the large reliance on private cars, "physical inactivity, obesity, death and injury from crashes, cardio- respiratory disease from air pollution, noise, community severance and climate change" [91,92]. Recent work quantifies, on the contrary, the economic benefits of reducing space devoted to cars in favor of green areas [93,94] and the negative outcomes of affluence.

It should also be noted that change in the end-use demand for services/sufficiency is not simply an issue of personal investment choices and behavioral changes at the individual level: sociological research indicates the need for enabling infrastructures and social frameworks [25,95]. This is also summarized in a UNEP report on "sustainable lifestyle" [96]: "A 'sustainable lifestyle' is a cluster of habits and patterns of behavior embedded in a society and facilitated by institutions, norms and infrastructures that frame individual choice, in order to minimize the use of natural resources and generation of wastes, while supporting fairness and prosperity for all". Hence, we explored in this section, institutions, norms and infrastructures that can foster energy sufficiency actions (sometimes overlapping with energy efficiency actions/technologies) at the building and district/city level showing that those two levels are strictly connected.

### 4.1. Designing (Spaces and Legislation) for Sufficiency

### 4.1.1. Comfort Scenarios, including Air Velocity and Ceiling Fans

A tendency to develop an architecture fully detached from the external environment and to aim at maintaining internal spaces strictly controlled in terms of temperature and humidity and with essentially zero air movement has dominated the second part of the XX century and the start of the XXI. This was paralleled by a rather narrow interpretation, by the construction and systems industry, of the then predominant comfort model, developed by Fanger [97], proposed for application in mechanically controlled environments. In reality the model allows for a rather large range of temperatures, also depending on clothing and chair insulation and activity levels, and does not mandate for a narrow range of humidity. Fanger states that "the influence of humidity [on comfort] is small" and presents calculations and graphs showing that a change of 1% R.H. produces changes of 1/100 to 1/1000 of a unit of PMV [97], while the comfort range spans from −1 to +1 in terms of PMV [97]. At the same time the adaptive comfort model has been developed based on a large body of data in real buildings [98,99] and was included in standards (EN 16798 [75] and ASHRAE 55 [76]) for non-mechanically conditioned spaces and for conditioned spaces when systems are turned off. The necessity for an extension of the PMV model was acknowledged also by Fanger [100]. The adaptive comfort model, which proposes a linear positive correlation of summer indoor comfort temperature with the average outdoor temperature in the previous week, allows for lower *energy needs* when compared to a restrictive interpretation and application of the PMV model [101], while providing comfort, based on an a very large database of measurements and surveys [99,102]. Finally the role of air velocity in providing comfort in the warm season at temperatures higher than calculated with the PMV formula has been confirmed in a long series of experiments and included in both EN 16798 and ASHRAE 55. By adopting higher air temperatures during warm seasons, building operators may reduce HVAC *energy use* by approximately 7–10 percent per degree Celsius of temperature increase [103].

However, in spite of the fact that results from over 35,000 occupant surveys contained in the ASHRAE Global Thermal Comfort Database [99] show that occupants prefer more air movement than they are currently experiencing in buildings, designers still have little guidance for designing rooms with ceiling fans (spacing, sizing and cooling effect) [104] and rarely ceiling fans are considered in the energy concept at the design stage and actually installed and coordinated with the lighting design. A new design and sizing tool has been created and made available by the Berkeley group [105], which also provided results and analysis from the largest study to date of air speeds generated by ceiling fans [106].

The possibility to apply (as a user of a building as much as a designer of a building) efficiency/sufficiency measures such as night ventilation in summer nights [107] and use ceiling fans during the day instead of (or to reduce use of) air-conditioning depends on explicit recognition at the regulation level of the following issues:

(1) In summer, the same level of thermal comfort, as measured, e.g., via the index predicted mean vote (PMV), can be achieved via various combinations of the physical parameters (operative temperature, relative humidity and air velocity), each scenario leading to different values of *energy need for cooling* and *energy need for dehumidification* (if any) [101,108].

(2) The choice of the comfort category (I, II or III according to EN 16798-1, formerly known as EN 15251, or A, B and C according to ASHRAE 55), which is aimed at the building design and/or controls that strongly affect *energy needs* [109].

(3) A number of research works show that comfort category I (A), which is the more energy demanding, cannot be perceived subjectively [110] and it is below the accuracy of measurements [111]. In the EU standard (EN 16798-1), category I (A) is reserved to buildings occupied by people with special needs (children, elderly, persons with disabilities, etc.), but it may nevertheless be perceived by designers and presented to clients/operators as the "best" condition.

(4) An important parameter affecting comfort in the warm season is the insulation level of clothing and of furniture, as e.g., office chairs (both measured in the unit clo and with indicative values reported e.g., in ISO 7730 [112]). Regulation and cultural norms may actively and explicitly promote the adoption of dressing codes where light clothing in summer is the norm rather than the exception (see e.g., the Cool Biz program in Japan [113]) and office furniture is chosen with low thermal insulation.

### 4.1.2. Using *Energy Needs* and *Total Primary Energy* as Indicators and Following the "Priority Order" Foreseen in the Standard EN-ISO 52000

A situation where all the actors involved in the development of efficiency/sufficiency measures in the field, regulators and policymakers use consistently the same set of physical concepts, definitions and nomenclature would ensure better final results in terms of comfort levels and energy use and would be a prerequisite for devising clear guidelines for design and construction focused on allowing sufficient behavior and operation. The necessity of using a unified nomenclature in legislation and regulation has been stated very explicitly in a report commissioned by the European Commission on ZEB definition [32] and has been supported in the scientific literature [34] and EU projects [31]. The Standard ISO EN 52000-1 [36] explains which indicators are needed (*energy needs for heating and cooling, total primary energy* use and *non-renewable primary energy* use) and why (see Section 2. Energy Efficiency). The indicators *energy needs for heating and cooling* and *total primary energy* do respond to the "energy efficiency first" principle and to the aim of quantifying the effect of sufficiency actions, while the indicator *non-renewable primary energy* use responds to the objective of "increasing the share of renewables". Lowering the value of those indicators via improvements of the building fabric, by providing a more uniform comfort in spaces and reducing the daily fluctuations of temperature facilitates the adoption of sufficiency actions by occupants of buildings (e.g., adapting clothing, prioritizing use of fans over air conditioning, etc.), which would not be possible in buildings with a poor quality of opaque and transparent envelope components.

### 4.1.3. Integration of Actions at the Building and District Level

Design of buildings as guided by building codes and city planning are still to a large extent dealt separately. On the contrary, sufficiency (and efficiency) actions in buildings are strongly connected with enabling/hindering conditions in cities. An overview of interactions between districts and buildings favoring efficiency and sufficiency actions and the necessary supporting urban design and regulation is presented in Table 2.

**Table 2.** Interactions between districts and buildings favoring efficiency and sufficiency actions and supporting urban design choices and regulation.

| Sufficiency Actions in Buildings→ | Summer Night Ventilation and Ceiling Fans Rather Than Air Conditioning | Summer Night Ventilation Rather Than Air Conditioning | Adequate m² per Capita Floor Space | Adopt "Sufficient" Mobility Modes: Bicycle, Walk, Public Transport | Line Drying and Water/Hot Water Saving |
|---|---|---|---|---|---|
| In order to perform sufficiency actions, inhabitants would need→ | Silence at night, clean air | External air temperature < 20 °C at night | Pleasant common indoor/outdoor spaces (shared guest rooms, music rooms, office space, playing spaces for children, etc.) to reduce the need for individual volumes | Easy access to services, schools, work and coworking spaces, equally distributed in the city; independence of movement for children and elders | Well-designed spaces for line-drying, installed water saving devices. Comfortable showers in place of bathtubs |
| Presently cities create constraints→ | Noise, mainly from cars and motorcycles. PM10, PM2.5 pollution and other air contaminants | Asphalt, city canyons | Inhospitable districts, obligation for car parking spaces at buildings and free car parking on streets | Distance between functions, unacceptable risks for cyclists, pedestrians and persons with disabilities | Dust in air |
| Cities should offer enabling conditions→ | Car-free residential districts and zones at 20 or 30 km/h | White/cool surfaces. Geometries facilitating air movement. Water surfaces and urban vegetation | Walkable, cyclable districts, green spaces, spaces for playing and spaces in the building for common activities | Equitable access to street space and equal access to various transportation modes | Information campaigns on water saving devices and on the high quality of drinking water from the tap |
| Legislation and Regulation shouldaddress→ | Objective and adequate temperature and humidity set-points in regulation. Limitations to car number and to speed limits to 20–30 km/h | Mandatory white/cool surfaces, mandatory external solar protections (as, e.g., in Switzerland) | Minimum requirements of green spaces and of common spaces for meetings | EPBD (and national build codes): mandatory protected spaces inside buildings for bicycles, wheelchairs and strollers | Mandatory spaces for line drying, mandatory labeling of low-flow water devices, mandatory showers rather or in addition to bathtubs (with access at the same level of the floor for easy access by aging population) |

Natural Ventilation and Ventilative Cooling Need Silence and Clean Air; Interactions with Mobility Planning

In the district of Florés Malacca, a group of buildings offers a recent example [114] of holistic planning of buildings and districts. Orientation of buildings and shape and openings of each apartments take advantage of dominant Alisee wind for achieving cross ventilation and cooling the building fabric at night. Ceiling fans are an integral part of the comfort concept. The presence of cars has been limited to an underground parking in order to achieve an acoustically quiet environment allowing the opening of windows for natural ventilation for a large fraction of time without acoustic discomfort and degradation of indoor air quality (IAQ) by external pollution.

Recent actions in large cities (e.g., Paris) aimed at creating opportunities for walking and cycling (in parallel to limiting individual motorized vehicles use and speed) and increasing green areas might allow for better opportunities also in the use of buildings, as in the previous example. Solar protection of streets and small squares is relatively common in some towns in the South of Spain and Portugal and by lowering the air temperature in the street canyon might allow for better conditions for night summer ventilation. The use of spaces for introducing vegetation and low solar absorptance surfaces, if practiced at a large scale, can reduce the heat island effect [115] and maintain the potential for using night ventilation as an effective passive cooling technique.

Analyses performed by the Department of Architecture and Urban Studies of Politecnico di Milano [116,117] show how a combination of diffused coworking spaces and safe biking infrastructure might allow a practice of "near working" coherent with the idea of "15 min city".

One important feature of some of the above described actions (e.g., shifting mode to walking and cycling for the short-mid range, relocating some type of work) is the fact that they can be implemented quite quickly; some of them have very low cost [118] and deploy effects in a very short time span [119], which are now available in order to limit emissions quickly enough to remain in the 1.5 °C carbon budget [120]. Just between 2011 and 2016 the world has wasted 200 out of 365 $GtCO_2$ of the available 1.5 °C $CO_2$ budget to 2100 [121].

Common Spaces and Dedicated Spaces for Line Drying and Bike Sheltering

Common spaces within buildings favor conviviality and cohesion and may reduce the need for excessively large private (conditioned) spaces. e.g., the Geneve Plan (2000 W) foresees the promotion of common spaces, the rationalization of empty private spaces via "la multifonctionnalité des espaces, le partage et le recyclage de l'espace, des équipements et des biens; la pratique du coworking, le télétravail, les coopératives d'habitation, les colocations". Common, car-free spaces outdoors (in particular green spaces) can also offer an important alternative to indoor spaces for many activities (e.g., children playing activities, elderly people physical exercise and social interaction) and, hence, deliver multiple benefits (reduced need of indoor-conditioned spaces, reduced pollution and noise and surfaces to be dedicated to heat island effect mitigation [122]). Green spaces between the buildings, on the roof and some facades, help reduce the air and surface radiant temperatures and hence create better conditions for effective summer night ventilation.

A provision for spaces adequate to line drying outdoors (on facades, balconies and roofs) and well designed for convenience and aesthetic can enable this practice, very relevant in terms of energy saving (drying a kilogram of clothes indoors or with a drying machine can be 3–5 times more energy expensive than washing it, due to the high value of the heat of vaporization of water).

Spaces for bike sheltering for all new buildings and renovations (in place of the current practice of imposing car-parking spaces) may support the modal shift and liberate public space for the common uses. The 2018 Energy Performance of Buildings Directive recast [11] calls MS to promote bike sheltering and holistic, coherent urban planning (article 8 and recital 28).

(Hot) Water Savings

Voluntary labeling schemes are available for taps and showers (e.g., the Water Efficiency Label, the Swiss Energy Label for Sanitary Fittings and the European Water Label scheme) and could potentially apply to almost all products used in domestic and non-domestic applications. A variety of water saving devices are available on the market: low flow shower heads and taps, asymmetric commands for avoiding unwanted use of hot water and devices with timers. Particularly effective are devices with the aeration mechanism that make use of the Venturi principle: while water passes through a restriction, its pressure is lowered below atmospheric pressure. Such a depression sucks in a certain amount of air, which mixes with the water, increasing its apparent volume. As an order of magnitude, the water consumption is reduced by 40–50% with respect to a traditional shower at parity of the "volume" feeling. The tendency to install showers rather than bathtubs may also be favored in new design and retrofits with the objectives of encouraging a lower use of water, of facilitating access to the elderly people (especially in places like EU where population is aging) and of saving conditioned space.

Drain Water Heat Recovery

Various heat exchanger typologies are available on the market to recover thermal energy from the water outflow of showers, bathtubs and sinks and preheat incoming water, e.g., in the gravity falling-film method, surface tension and gravity cause falling films of water to adhere to the inner wall of a vertical drainpipe, thus enabling a high rate of heat transfer. The cold water from the aqueduct passes through a coil that is tightly wrapped around the vertical drainpipe. This system can recover 45–65% of the available heat in the wastewater. In cases that a vertical pipe proves difficult to install, horizontal heat exchangers are available, with slightly lower efficacy. The system cost effectiveness is obviously higher in the cases of a large use of water as, e.g., in sport facilities [123].

## 5. The Case Study of the Chiaravalle District in Milan, Italy
### 5.1. Description of the Case Study

We present here a case study of transformation of an existing district into a PED where part of the measures have been already successfully realized with the retrofit of multiapartment social housing buildings and a part is under analysis/planning. We performed a "what if" analysis comparing two scenarios:

- Scenario (A) under which only active systems were upgraded/installed (heat pumps and PV) and there was no change in *energy needs/uses*;
- Scenario (B) where a series of efficiency measures and sufficiency enablers were implemented thus strongly reducing the *energy needs/uses*; in addition, active systems were upgraded/installed.

A strong focus was given to measures for summer comfort in the expectation of a further increase of temperature and exacerbation of the heat island effect. We compared the effects under both a yearly and seasonal (winter/summer) balanced approach and estimated the resulting land take necessary for achieving a zero (or positive) energy balance.

The Chiaravalle district is located in the southern periphery of Milan, Italy, at the border with the Parco Agricolo Sud Milano nature reserve. The size of the project area is around 330,000 m$^2$, comprising a monastery, public and private residential units, public offices, restaurants, hotels and public spaces (Figure 3).

**Figure 3.** Map of Chiaravalle district, Milan, Italy: in red the public housing blocks are highlighted. The analyses performed by the authors are referred to complex A but the results can be extended to complex B, because of their similarities.

Despite being annexed to Milan, Chiaravalle has maintained the character of an autonomous town. The district is separated from the urban agglomeration of the city by the park and this isolation is reinforced by the scarcity of public transportation to the centre. The district is undergoing renovation in the framework of several EU projects (e.g., EU-Gugle, SharingCities and SATO—self assessment towards optimization of building energy) aimed to rehabilitate the area starting from the deep energy retrofit and the energy optimization of four residential apartment blocks (highlighted in red in Figure 3).

They are public housing blocks, consisting of L-shaped buildings with four stories each. This kind of housing represents an important share of the national public building stock; most of these buildings were built in the 1970s and 1980s and have never been renovated, thus presenting both poor energy performance and serious IAQ problems. The needed retrofit has been carefully planned and executed without moving the occupants out of their apartments.

We focused the analysis on the complex A (Figure 3), but the results can be extended to the complex B, since the buildings are characterized by a similar shape and materials and have undergone the same type of retrofit. A number of similar buildings are present in the national building stock so this analysis may provide indications on the potential of deep retrofitted buildings for sufficiency, efficiency, flexibility and coverage of energy needs via on-site renewables.

The complex is made of two buildings, named Building 1 and 2 (Figure 4). They present a gross surface area equal to respectively 1797 m$^2$ and 2836 m$^2$, accounting for 66 residential units and an estimated population of 210 persons.

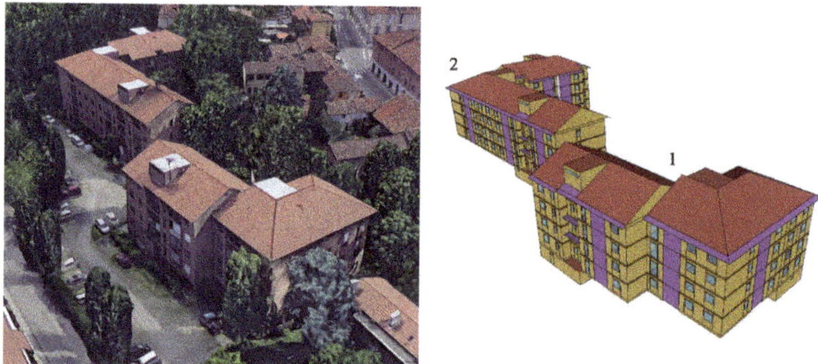

**Figure 4.** Aerial view and geometrical model of the public housing estate.

The building envelope is made of prefabricated concrete elements and presents a window/wall ratio of 14%. The pre-retrofit state included thin layers of thermal insulation material only in some areas, and presented low performance windows with no proper solar shading (only roller shutters that when operated blocked completely both daylight and ventilation). The centralized heating system used fuel oil as the energy carrier ($\eta = 0.7$), whereas each apartment was equipped with a local boiler for DHW generation ($\eta = 0.7$), using natural gas as the energy carrier. Natural gas was used also for cooking, while all the other energy uses relied on electrical energy, supplied by the national grid.

The retrofit aimed at improving first of all the energy performance of the building envelope: substantial exterior insulation of the opaque elements including walls, roof and exposed ground floor slab, extremely detailed reduction of thermal bridges, low-e double pane glazing and window frames with thermal break and exterior solar blinds (Figure 5). Table 3 shows the physical characteristics of the building fabric, before and after the deep energy retrofit.

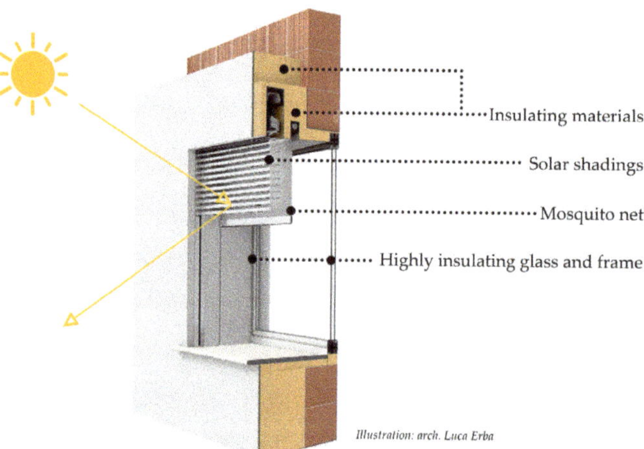


**Figure 5.** Window-wall detail after the deep energy retrofit: the new windows, characterized by highly insulating glass and frame, are protected by movable solar blinds and mosquito nets, which create a barrier for insects with a small reduction of the air flow when natural ventilation is used.

**Table 3.** Physical characteristics of the building, before and after the renovation.

| | Before Renovation | After Renovation |
| --- | --- | --- |
| Thermal transmittance of opaque vertical structures U (W/(m$^2$K)) | 1.15 | 0.13 |
| Thermal transmittance of the ceiling under the uninhabitable attic U (W/(m$^2$K)) | 3.00 | 0.15 |
| Thermal transmittance of the pilotis supported slab U (W/(m$^2$K)) | 2.40 | 0.17 |
| Thermal transmittance of glass panes U (W/(m$^2$K)) | 3.00 | 1.42 |
| Thermal transmittance of the window frames U (W/(m$^2$K)) | 5.00 | 1.60 |
| Total solar transmittance of glass panes (%) | 0.75 | 0.52 |

After the envelope renovation design, the project focused on improvements of the building systems, as follows:

- Installation of a centralized generation system for heating and DHW based on water-to-water heat pump technology (92 kW and a seasonal coefficient of performance (SCOP) of 2.7 according to standard EN 14825 [124]);
- Installation of thermostatic valves on each radiator;
- Installation of a centralized mechanical ventilation system with heat recovery and bypass (to allow for free cooling in summer and mid seasons) and an average specific fan power of 2 kW/(m$^3$/s);
- Installation of LED lamps for common area lighting.

Finally, according to the climatic conditions and to the available roof surface, a PV system for the exploitation of renewable energy sources was designed and installed.

The analysis of the building behavior was performed through the energy model that was developed in an EnergyPlus (Illinois–CA, USA) simulation environment with a high number of thermal zones, in order to be able to assess the behavior of the individual apartments including the effect of the orientation. The geometric modeling was realized on the basis of laser scanning topographic surveys, validated and corrected by on-site inspections and verifications. The model considers the presence of thermal bridges, characterized by the calculated linear thermal transmittance value before and after the retrofit. Building 1 was subdivided into 41 thermal zones and Building 2 in 66 thermal zones. In particular, twenty-two of the twenty-four apartments of Building 1 were modeled as individual thermal zones; the remaining two apartments, which accommodated an advanced indoor environmental monitoring equipment for assessing indoor comfort conditions, were modeled considering a thermal zone for each single room (for a total of about seven thermal zones per apartment). Finally, six thermal zones were dedicated to unheated environments: staircases, basement and attic floor. In Building 2, thirty-nine of the forty-two apartments were modeled as individual thermal zones; the remaining three were modeled considering a thermal zone for each single room (for about seven thermal zones per apartment), whereas eight thermal zones were used for unheated environments.

For the calculation of the *energy needs*, the heating system was characterized in EnergyPlus by an ideal system able to maintain a temperature of 20 °C during the heating season that is defined according to Italian national regulations from 15 October to 14 April, for the considered climatic zone. During the cooling season, from 15 April to 14 October, an ideal active cooling system able to maintain an indoor set-point temperature of 26 °C was simulated. The mechanical ventilation system was modeled considering 0.5 air changes per hour (ACH) with night attenuation (22:00–6:00) equal to 0.25 ACH. In the pre-retrofit model, an air infiltration value of 0.5 ACH was set for the apartments and staircase units, whereas it was set equal to 1 ACH for unheated areas. In the post-retrofit model, air infiltration was reduced to 0.05 ACH in the apartments and to 0.5 ACH for staircases. After a sensitivity analysis, it was decided to activate the shading devices in the post-retrofit whenever the solar irradiance level exceeded 200 W/m$^2$, simulating an "average" occupant behavior. In the pre-retrofit model, no shading device was considered; only rolling shutters were applied at night-time. To simulate the presence of people and therefore to improve the estimation of internal gains, a schedule based on measured electric consumption was created (the input values were used after a careful analysis on the quality of existing electrical energy data). Further analyses based on data-driven procedures were performed to derive even more detailed occupancy and occupant-related load profiles and are presented in [125].

*5.2. Assessing Energy Efficiency Improvements*

The effectiveness of the renovation actions was evaluated calculating the *energy needs for heating and cooling* per net conditioned floor area before and after the deep energy retrofit. The simulations were performed considering as the yearly outdoor weather dataset the typical meteorological year (TMY) file of the location, obtained through the TMY tool developed and updated by the Joint Research Centre of the European Commission.

The weather dataset was characterized by 2593 heating degree days (HDD) and by 69 cooling degree days (CDD), calculated according to ISO 15927-6:2007 [126]. Figures 6–8 show respectively the hourly outdoor air temperature, the monthly global horizontal solar irradiation and the frequency distribution of wind speed and direction in Chiaravalle, Milan.

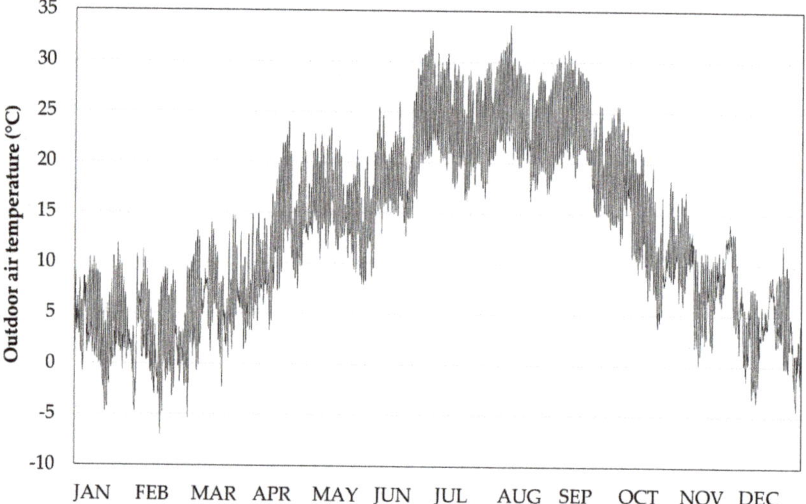



**Figure 6.** Hourly outdoor air temperature distribution in Chiaravalle, Milan (latitude: 45.417, longitude: 9.214, height: 105 m).

The outputs of the dynamic simulations show, as a result of the deep retrofit, a reduction of the yearly *energy need for heating* from 147.4 to 16.6 kWh/m$^2_{net}$ and of the yearly *energy need for cooling* from 19.6 to 9.1 kWh/m$^2_{net}$. In summer, the application of natural night ventilation and the use of ceiling fans for achieving comfort at higher air temperature was expected to further significantly reduce the *energy needs for cooling*. Conservatively we assumed a remaining level of *energy need for cooling* at 7 kWh/m$^2$y.

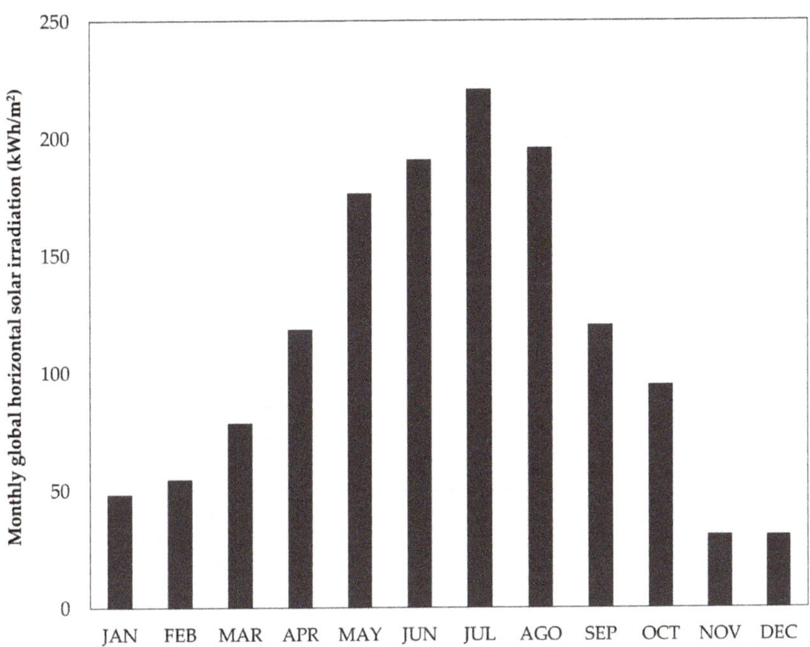

**Figure 7.** Monthly global horizontal solar irradiation in Chiaravalle, Milan (latitude: 45.417, longitude: 9.214, height: 105 m).

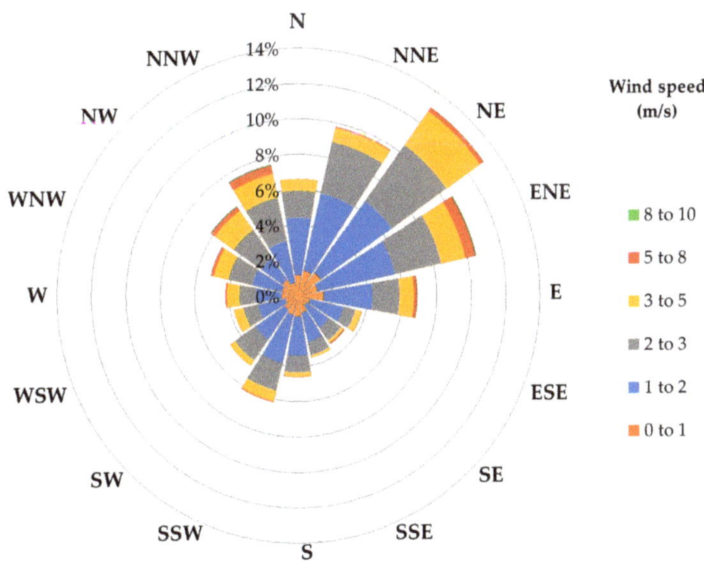


**Figure 8.** Frequency distribution of wind speed and direction during the year in Chiaravalle, Milan (latitude: 45.417, longitude: 9.214, height: 105 m).

*5.3. Assessing the Energy Flexibility Potential after the Renovation*

To evaluate the storage potential of the thermal mass during winter [127], after the renovation, in our simulations we brought the conditioned space to an operative temperature of 24.1 °C. This temperature corresponds to the upper value of the comfort range of category II using the Fanger PMV model, when assuming typical indoor winter clothing (1 clo), metabolic activity corresponding to office work (1.2 met), low air velocity (0.1 m/s) and relative humidity (40%) [75].

This temperature was maintained in the simulation for respectively 1, 2, 3, 4 and 5 days. Afterwards the heating system was turned off. In order to eliminate climate variability from the calculation and focus the analysis on the effect of the thermal capacity plus thermal insulation in delivering flexibility, we selected an "average winter day" (Figures 9 and 10), which during a set of simulations was cyclically repeated.

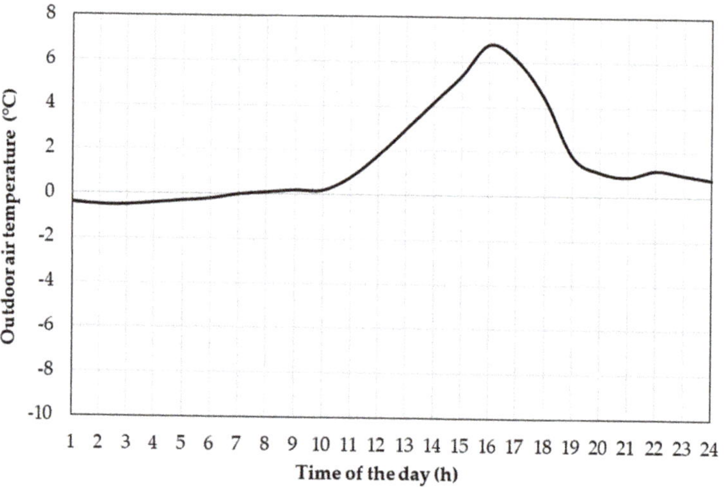

**Figure 9.** Outdoor air temperature during an "average winter day".

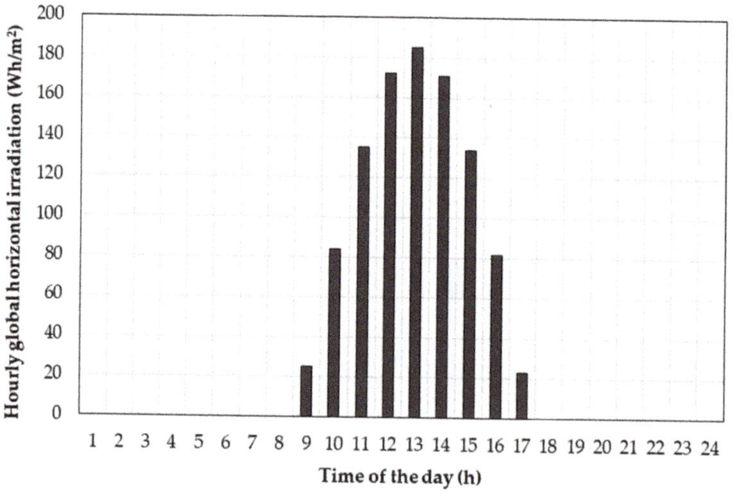

**Figure 10.** Hourly global horizontal irradiation during an "average winter day".

The time interval during which the space remained in the comfort zone (24.1–19.5 °C), under a climate given by the cyclic repetition of an "average winter day", was different for each of the five cases, but the difference was limited. By heating up (within the comfort zone) the envelope for 1 day, the conditioned space will remain in the comfort zone after turning off the heating system for approximately 4 days (96 h, as shown in Figure 11). By heating up the envelope for 2 days, the conditioned space remained in the comfort zone after turning off the heating for more than 5 days (more than 120 h, as shown in Figure 11). A further increase in the time interval during which the heating was kept on produced marginal results showing that it was possible to activate a large part of the thermal storage potential by keeping the heating system turned on for just one day.

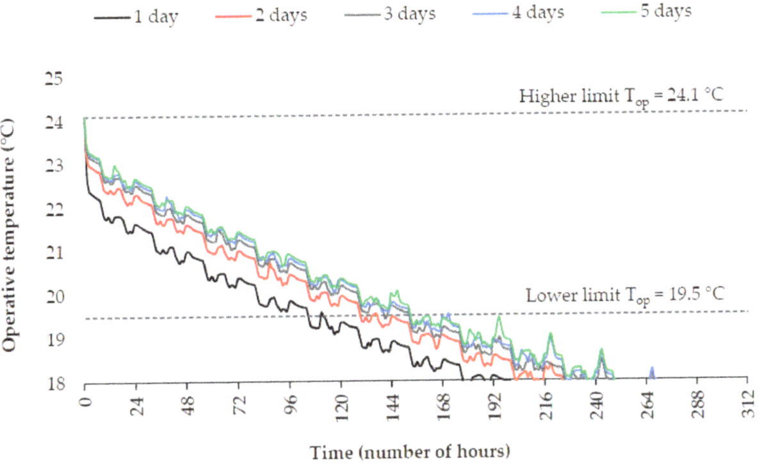


**Figure 11.** Variation over time of operative temperature in a reference indoor heated space as a function of the number of days (1–5) during which the heating system was kept on (before being turned off), with external conditions given by the cyclic repetition of an "average winter day".

The hypothesis underlying the study is that the increased external thermal insulation of walls, roofs and basements, and the heat recovery on ventilation, considerably extends the time interval during which a building will maintain indoor conditions in the comfort range. This fact will make it possible to: (i) coordinate the demand with the supply of local energy, by removing the current rigidity of energy demand from buildings and thus allow them to "ask for" energy precisely when it is available from local sources (renewable or recovered energy) or to exchange energy with other buildings in a flexible way; (ii) exploit moments of supply overabundance of renewable energy on the grid by making available energy storage capabilities (in the form of thermal capacity of the building fabric) when such moments occur; (iii) manage conditions of energy supply shortage by attenuating the peak power demand on the grid or district heating network (peak shaving, demand response, potential participation in the capacity market creating added value that is in addition to the value associated with energy savings and increased comfort).

In the case of a series of adverse days (night temperature dropping to $-7\,°C$ and maximum hourly irradiation reduced to $100\ \mathrm{Wh/m^2}$), the time of permanence in comfort after one day of "thermal charge" reduced slightly, to 70 h, e.g., about 3 days.

Figure 12 shows that the substitution of windows and doors was not, by itself, enough to significantly modify the thermal dynamics of a building with a window/wall ratio typical of existing residential buildings in Italy. Thermal insulation of the opaque parts of the building fabric to the level of quality taken under consideration in this case (conductivity in the order of 0.035–0.040 W/(m·K) and a 25 cm thickness of external insulation) was indispensable for obtaining building flexibility (in addition to saving energy by reducing the *energy need for heating*). Obviously, limiting the renovation to just the substitution of the thermal energy generation system, without any intervention on the building fabric, would have no effect on the flexibility.

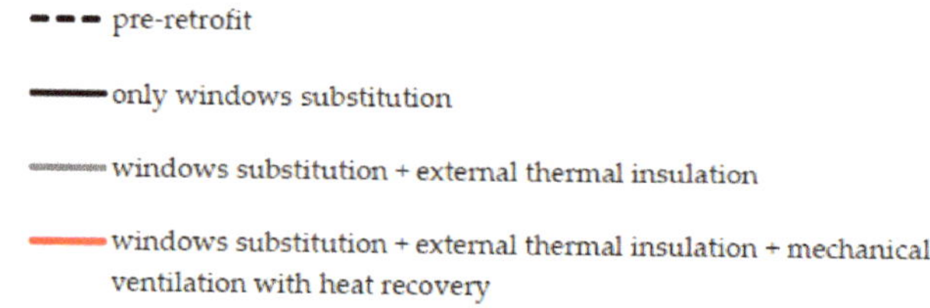


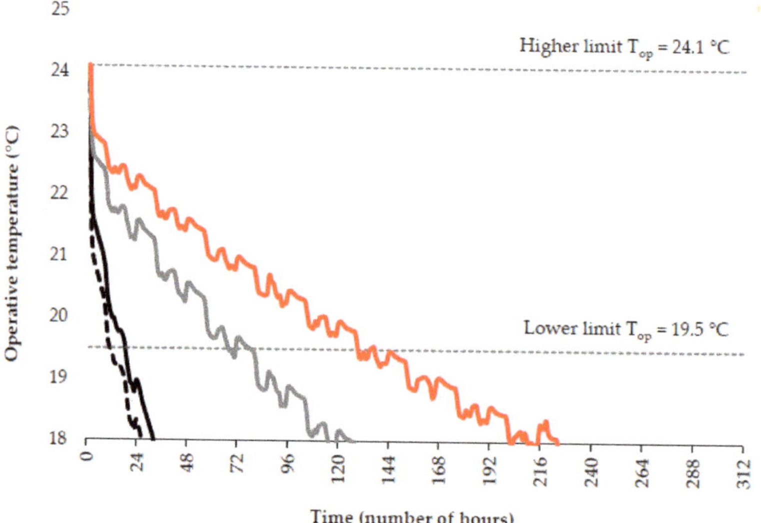


**Figure 12.** Variation over time of the operative temperature in a reference heated space as a function of the retrofit measures undertaken, after two days during which the heating system was kept on (before being turned off) and with the cyclic repetition of an "average winter day".

Considering the situation before renovation of the building fabric, even delivering heat for one day with a set-point temperature of 24 °C, once the heating system was turned off the building will only remain in the comfort range for about ten hours. The effect of a potential shift of the demand was limited and very costly in terms of energy dissipation. In fact, most existing buildings behaved like short-circuited (thermal) batteries.

Thermal insulation of the building fabric, possibly with the addition of a mechanical ventilation system with heat recovery, proves to be an indispensable condition to enable the building to be flexible with respect to the moment in time when energy is required. Most of the current buildings were "rigid" in this respect and in winter they all needed energy in the same morning time slot after the reduction of the set-point temperature at night.

"Storage of Coolness" in the Thermal Mass

The achieved "physical" flexibility will be exploited with the support of the H2020 project SATO. *Energy use* and state of various systems (heat pumps, mechanical ventilation and PV panels on the roof) and indoor condition of some representative apartments will be continuously monitored and the optimization procedure will provide instructions to the building manager and occupants on how to take profit of the local availability of renewable energy in order to "thermally charge" the structures and avoid energy use at times of low or zero availability of RES energy, taking into account also current and forecasted weather conditions. This "Building as Battery" concept will allow one to combine in the best way "physical" flexibility and control and optimization, where the storage cost will be essentially zero since the insulation that avoids the "batteries" to quickly discharge into the environment is paid back already and in a short time via the energy savings.

### 5.4. Enabling Energy Sufficiency Actions

As we discussed in previous sections, sufficiency actions could be hindered or enabled by physical and regulatory infrastructures. In the case of the considered Chiaravalle buildings a number of enabling features are already installed, others are in the phase of evaluation and planning.

### 5.4.1. Enablers Already Approved/Installed

The relatively ample balconies will enjoy external movable shading devices and external movable insect screens (Figure 5); this creates an external area where to spend time in the evening in the summer and mid seasons and allows for night ventilation, complementing the fact that almost all apartments have openings in opposite facades of the building, hence allowing for cross ventilation and heat flushing of the thermal mass. The thick external insulation contributes to greatly smoothing and delaying the heat wave crossing the walls and ceiling, hence reducing mean radiant temperature of the rooms, in cooperation with external solar protections at each opening and cross night ventilation. Some of the occupants spontaneously reported their appreciation for the substantial reduction of the temperature of the indoor surfaces in summer compared to the pre-retrofit situation.

These improvements in comfort by passive means might enable occupants to adopt sufficiency behaviors such as moving to the fresher part of the space, adapting clothing, and avoiding the installation/use of air conditioning units.

### 5.4.2. Enablers under Analysis/Planning

- Offering support to families for the choice and installation of efficient ceiling fans, which would allow it to deliver at 28–29 °C, the same summer comfort level as at 25–26 °C according to a large number of experiments and surveys in many world locations and consolidated in the new version of comfort standards (EN 16798 and ASHRAE 55).
- Installation of a water tap on the balconies for easing the installation and use of simple sprinkles to add evaporative cooling in extreme days.
- Offering support (or direct installation) of low flow shower heads and heat recovery on drainage water.
- National funding is available for the above improvements (both for devices and installation cost) up to 65% via tax rebates.
- Installation of well designed, comfortable to use devices for line drying outdoor, as is traditional in many parts of Italy.
- Creation of shelters for bikes, cargo bikes and strollers in some of the rooms on the ground floor.
- Creation of a bike path, separated from the road, along the path of a disabled train line, to connect the Chiaravalle district to the public transit hub of Rogoredo (where it is possible to take long distance trains, urban trains and metro, buses; the Duomo, center of the city, can be joined from Rogoredo in 12 min by metro).
- Reduction of the velocity limit to 20 km/h in the whole district and the availability of alternatives to private cars (shared electric/cargo bikes and trolleys and an automated small bus taking profit of the abandoned rail connecting Chiaravalle to Rogoredo to allow more flexible rides in addition to the present bus line).

The whole measures for enabling a significant modal shift from private cars to biking and public transit is supposed to reduce the use of a private car from 12,000 to 5000 km per household. In both scenarios it is assumed that a small electric car per family was used, with a performance of 18.6 kWh per 100 km [128].

### 5.5. Yearly and Seasonal Energy Balance of the District

Based on the data from the literature and our own analyses described in the previous sections of the paper, we drew an energy balance of a representative unit footprint ($1\ m^2$) of the buildings, in terms of orders of magnitude. The analysis reported in Table 4 was performed for the entire year and for "winter" (October to March) and "summer" (April to September). We assumed an average height of the building of four stories, which is representative of the district. A number of input data were taken from national statistics, such as average electric *energy use*/household, flat average size, in order to be able to generalize the result to other situations. Sources of data are indicated in the table; specific calculations performed by the authors are briefly described in the notes and were based on the energy efficiency and sufficiency options analyzed in the previous sections. The objective was to identify the key areas and to draw general conclusions on the feasibility of a net (or positive) energy balance and the land take impact of this objective in the two scenarios:

(A)  In which no action was taken to reduce the *energy needs* and only supply was improved, by installing a heat pump for heating/cooling and DHW and PV on the rooftop (consistency of supply);

(B)  In which *energy needs* were reduced by sufficiency and efficiency measures and supply was improved by installing a heat pump and PV on the rooftop and part of the facades (sufficiency, efficiency and consistency of supply).

**Table 4.** Overall energy balance over a year, "winter" and "summer".

| | | SI Units | Scenario A (Whole Year) | Ref. | Scenario A ("Winter") | Scenario A ("Summer") | Scenario B (Whole Year) | Ref | Scenario B ("Winter") | Scenario B ("Summer") |
|---|---|---|---|---|---|---|---|---|---|---|
| | | | **Scenario A (Only Consistency of Supply)** | | | | **Scenario B (Sufficiency, Efficiency and Consistency of Supply)** | | | |
| Reference building | Apartment average floor area | $m^2$ | 105.9 | [129] | | | 105.9 | [129] | | |
| | SEER, SCOP | | 2.7 | [124] | | | 2.7 | [124] | | |
| | Ventilated volume | $m^3/m^2$ | | | | | 2.7 | (a) | | |
| | Air changes/h | vol/h | | | | | 0.5 | (b) | | |
| | Air volume flow rate | $m^3/s$ | | | | | 0.00038 | * | | |
| | SFP | $kW/(m^3/s)$ | | | | | 2 | (c) | | |
| | Mechanical power for ventilation | $kW/m^2$ | | | | | 0.00083 | * | | |
| | Average number of hours of ventilation (15/10 to 15/04) | h | Not active | | | | 4320 | (d) | | |
| Energy needs | *Energy need for heating* | $kWh_{thermal}/m^2/y$ | 147.4 | (e) [16] | | | 16.6 | (e) | | |
| | *Energy need for hot water* | $kWh_{thermal}/m^2/y$ | 20.0 | [16,130] | | | 12.0 | (f) | | |
| | *Energy need for cooling* | $kWh_{thermal}/m^2/y$ | 19.6 | (e) | | | 7.0 | (e, f) | | |
| Energy uses | *Energy use* for heating | $kWh_{electric}/m^2/y$ | 54.6 | * | 54.6 | | 6.1 | * | 6.1 | |
| | *Energy use* for hot water | $kWh_{electric}/m^2/y$ | 7.4 | * | 4.1 | 3.3 | 4.4 | * | 2.4 | 2.0 |
| | *Energy use* for cooling | $kWh_{electric}/m^2/y$ | 7.3 | * | | 7.3 | 2.6 | * | | 2.6 |
| | Average electricity use in apartments (for appliances, lighting, ICT, etc.) | $kWh_{electric}/y$ | 2870 | [131] | | | 1800 | [132] | | |
| | Electricity use in apartments (for appliances, lighting, ICT, etc.) | $kWh_{electric}/m^2/y$ | 27.1 | * | 15.1 | 12.0 | 17.0 | * | 9.5 | 7.5 |

**Table 4.** *Cont.*

| | SI Units | Scenario A (Whole Year) | Ref. | Scenario A ("Winter") | Scenario A ("Summer") | Scenario B (Whole Year) | Ref | Scenario B ("Winter") | Scenario B ("Summer") |
|---|---|---|---|---|---|---|---|---|---|
| | | **Scenario A (Only Consistency of Supply)** | | | | **Scenario B (Sufficiency, Efficiency and Consistency of Supply)** | | | |
| *Energy use* for cooking | $kWh_{electric}/m^2/y$ | 3.9 | [133,134] | 2.3 | 1.6 | 3.3 | [133,134] | 2.0 | 1.3 |
| *Energy use* by ceiling fans | $kWh_{electric}/m^2/y$ | Not installed | | | 0 | 0.2 | [105] | | 0.2 |
| *Energy use* by elevators | $kWh_{electric}/m^2/y$ | 3.0 | [135] | 1.5 | 1.5 | 1.8 | [135] | 0.9 | 0.9 |
| *Energy use* for lighting in common areas | $kWh_{electric}/m^2/y$ | 1.2 | [136] | 0.7 | 0.5 | 0.6 | [136] | 0.4 | 0.2 |
| *Energy use* for mechanical ventilation | $kWh_{electric}/m^2/y$ | Not installed | | 0 | 0 | 3.6 | * | 3.6 | 0 |
| Space travelled with electric private vehicle/household | km | 11,885 | [137] | | | 5000 | (f) | | |
| Energy use per distance traveled of a small sized electric car | $kWh_{electric}/km$ | 0.186 | [128] | | | 0.186 | [128] | | |
| Total use with private vehicle | $kWh_{electric}/m^2/y$ | 20.9 | * | 12.2 | 8.7 | 8.8 | * | 5.1 | 3.7 |
| Public transport use | $kWh_{electric}/m^2/y$ | 2.0 | [138] | 1.0 | 1.0 | 5.0 | [138] | 2.5 | 2.5 |
| Total energy use (delivered energy) per unit conditioned floor area | $kWh_{electric}/m^2/y$ | 127.3 | * | 91.6 | 35.8 | 53.5 | * | 32.6 | 20.9 |
| Average number of stories | | 4 | (c) | | | 4 | (c) | | |
| Total energy use (delivered energy) per unit footprint area of the building | $kWh_{electric}/m^2/y$ | 509.4 | * | 366.2 | 143.2 | 214.0 | * | 130.3 | 83.7 |
| Generation by PV on roof per unit footprint area of the building | $kWh_{electric}/m^2/y$ | 120.0 | (g) | 39.0 | 81.0 | 120.0 | (g) | 39.0 | 81.0 |
| Generation by PV on facades per unit footprint area of the building | $kWh_{electric}/m^2/y$ | Not installed | | 0 | 0 | 95.0 | [139] | 30.9 | 64.1 |
| **Overall energy balance** | $kWh_{electric}/m^2/y$ | −389.4 | * | −327.2 | −62.2 | 1.0 | * | −60.4 | 61.5 |
| **Land take necessary to achieve a zero energy balance** | $m^2_{land}/m^2_{footprint}$ | −3.2 | * | −8.4 | −0.8 | 0 | * | −1.5 | 0 |

(Row groups at left margin: *Mobility* covers the mobility rows; *RES* covers the generation rows.)

(a)—An average height equal to 2.7 m is considered. (b)—Air change per hour defined according to the Italian standard. (c)—Values according to the technical documentation of the project, see Section 5.1. (d)—The number of hours during which the mechanical ventilation is active is calculated considering full utilization (24/7) during the heating season (from 15/10 to 14/04 in the climatic zone E). (e)—The values of *energy needs* have been estimated through the dynamic energy simulations realized for the case study (see Section 5.2). (f)—The values reported take into account energy sufficiency actions described in Section 4.1 and 5.4. (g)—These values were calculated via PVGIS simulations assuming: solar radiation database PVGIS-SARAH; PV technology crystalline silicon; installed peak PV power of 0.205 kWp/m²; system loss of 14%; slope of 35°; azimuth of +45° and −45°. For each m² of footprint 0.5 m² of PV panels can be installed, taking into account the geometry of the roof and the orientation of the building. * All these values were calculated using the values shown in the text, defined on the base of literature or simulations (see letters a–g).

## 6. Discussion

The challenge of defining and realizing positive energy districts is made complex by various factors, in particular if the balance is performed over the entire year rather than on smaller time intervals. For example, if one assumes the aim of a net yearly zero (or positive) energy balance between exports to and imports from the grid, using any energy storage creates a penalty by increasing *energy use*, due to unavoidable energy losses of the storage. Hence a conflict arises with the objective of maximizing the self-use of on-site generated renewable energy at the time when it is available, that is with the objective of being "flexible" about the time when the district uses energy, either for direct services or for storage and delayed use.

Both at the district and building level, another general critique holds with respect to an energy balance performed over a year when compensation is allowed between different time steps: it allows one to compensate for continued fossil use in some moments (e.g., winter) with overproduction of RES at other times (e.g., summer). This obviously does not lead to zero emissions and/or requires large interseasonal storage locally or somewhere in the network, with associated land occupation, energy losses and use of materials, embedded energy and related costs.

In the case of buildings, some MS, such as Italy and Spain, have chosen in their Energy Performance Certificate to perform the balance with only partial compensation, meaning that compensation happens within months but not between different months. Additionally, in the future even lower time periods for compensation might be chosen in case more detailed calculation methods would be adopted for the energy certification (e.g., when hourly calculation methods would be adopted), rather than the most current calculation method, based on monthly average weather data. A similar accounting method to limit the negative effects of the simplistic yearly balance should be adopted also when performing the energy balance of a district.

An additional difficulty arises due to urban density when comparing the load with the available solar production per unit area of the buildings footprint. To show the order of magnitude of the mismatch we performed the assessment presented in Section 5.5. We performed the energy balance at a yearly level and further detailed it into the "winter semester" (October to March) and "summer semester" (April to September) in the two scenarios:

(A) In which no action was taken to reduce the *energy needs* and only supply was improved, by installing a heat pump for heating/cooling and DHW and PV on the rooftop (consistency of supply);

(B) In which *energy needs* were reduced by sufficiency and efficiency measures and supply was improved by installing a heat pump and PV on the rooftop and part of the facades (sufficiency, efficiency and consistency of supply).

Aiming at a yearly net zero energy balance in scenario A, the *energy use* was so high that, in order to generate energy with additional PV installed on the ground, it would be necessary to take a land approximately 3.2 times larger than the footprint of the buildings, e.g., by subtracting that land to the adjacent park. However, obviously the balance was made over one year, so this accounting was virtual rather than physical. Even adding a PV plant 3.2 times larger than the district would only bring to zero the "paper" yearly accounting. In physical terms there will be an overproduction in summer and a portion of uses in winter not covered by the PV generation. Hence the buildings would still rely on fossil fuels and continue to generate emissions.

By performing an analysis where the aim is to achieve a zero energy balance per semester ("winter" and summer"), which is making the "paper" balance just a bit closer to physical reality, the land required in scenario A to achieve zero balance in winter was 8.4 times larger than the footprint of the buildings, assuming that no interseasonal storage was available. Achieving a zero energy balance month by month would require even more land.

In scenario B, where sufficiency and efficiency allowed for a significant reduction of *energy needs* and uses, the addition of a reasonable amount of PV on facades would allow one to obtain on paper a yearly zero energy balance and avoid any additional land use. Obviously, when performing the balance per semester the result was different. A zero energy balance in winter still required additional surface for installation of PV, but this time the surface was 1.5 times the footprint of the buildings and one could imagine to find this surface, e.g., via structures suspended over the parking lots serving the Abbazia, a fraction of the area of the water treatment plant, portions of the adjacent provincial road and train rails in such a way to avoid the need to take land from the surrounding agricultural and park area.

Even in this case obviously there would remain additional penalties for the losses due to daily and seasonal mismatch of generation and use. However, the scale of the challenge and of the use of material and land was reduced by a factor of about 6. In fact the situation in case A would be even more unfavorable, since demand in that scenario was highly inflexible, hence in the physical reality the generated PV will only partially be used onsite due to mismatch between the time of supply and time of demand. In case B the flexibility of demand will be much higher and a better match between generation and demand will be possible. Taking into account the negative effect of inflexibility, scenario A would probably need a land take of at least an order of magnitude higher than scenario B.

In conclusion the problem of land requirement might be greatly reduced when applying efficiency and sufficiency measures, which would drastically reduce *energy needs* and increase flexibility.

Similarly for meeting the challenge to provide storage at various time scales, from the daily to the interseasonal, a strong reduction of *energy needs for heating and cooling* via efficiency techniques and physical and regulatory frameworks that enable low-energy lifestyles (sufficiency) might prove decisive since this would reduce the size of required storage and the connected embedded energy and energy losses.

Further research work is required to explore in more detail the potential offered by the combination of a) increased flexibility in demand offered by improvements of the building fabric and heat recovery on ventilation and b) controls that optimize the use of this flexibility based on forecasts of weather, use conditions and renewable generation, e.g., in the following days up to one week. The authors plan to contribute to this further analysis within the ongoing H2020 SATO project.

**Author Contributions:** Conceptualization, S.E. and L.P.; methodology, S.E. and L.P.; software, S.E.; formal analysis, S.E. and L.P.; writing—original draft preparation, S.E. and L.P.; writing—review and editing, S.E. and L.P.; visualization, S.E.; funding acquisition, S.E. and L.P. All authors have read and agreed to the published version of the manuscript.

**Funding:** The study was partly developed within the framework of the project SATO (Self Assessment Towards Optimization of Building Energy), which has received funding from the European Union's Horizon 2020 research and innovation programme under grant agreement No 957128. This study will also contribute to the project FULFILL (Fundamental Decarbonisation Through Sufficiency By Lifestyle Changes), which has received funding from the European Union's Horizon 2020 research and innovation programme under grant agreement No 101003656, as per the potential role of sufficiency.

**Acknowledgments:** We acknowledge help in data collection and useful discussions with Arch. S. Bardeschi and F. Manzoni of Municipality of Milano, A. Roscetti and A. Barbieri from eERG-Polimi and stimulating discussions within EERA (European Energy Research Alliance).

**Conflicts of Interest:** The authors declare no conflict of interest.

# References

1. United Nations. Resolution Adopted by the General Assembly on 25 September 2015-Transforming Our World: The 2030 Agenda for Sustainable Development. Available online: https://www.un.org/ga/search/view_doc.asp?symbol=A/RES/70/1&Lang=E (accessed on 3 July 2021).
2. Grubler, A.; Bai, X.; Buettner, T.; Dhakal, S.; Fisk, D.J.; Ichinose, T.; Keirstead, J.E.; Sammer, G.; Satterthwaite, D.; Schulz, N.B.; et al. Urban Energy Systems. In *Global Energy Assessment (GEA)*; Cambridge University Press (CUP): Cambridge, UK, 2012; pp. 1307–1400.
3. Schraven, D.; Joss, S.; de Jong, M. Past, present, future: Engagement with sustainable urban development through 35 city labels in the scientific literature 1990–2019. *J. Clean. Prod.* **2021**, *292*, 125924. [CrossRef]
4. Moreno, C.; Allam, Z.; Chabaud, D.; Gall, C.; Pratlong, F. Introducing the "15-Minute City": Sustainability, Resilience and Place Identity in Future Post-Pandemic Cities. *Smart Cities* **2021**, *4*, 93–111. [CrossRef]

5.  Yeung, P. How "15-Minute Cities" Will Change the Way We Socialise. Available online: https://www.bbc.com/worklife/article/20201214-how-15-minute-cities-will-change-the-way-we-socialise (accessed on 30 July 2021).

6.  Maestosi, P.C.; Andreucci, M.B.; Civiero, P. Sustainable Urban Areas for 2030 in a Post-COVID-19 Scenario: Focus on Innovative Research and Funding Frameworks to Boost Transition towards 100 Positive Energy Districts and 100 Climate-Neutral Cities. *Energies* **2021**, *14*, 216. [CrossRef]

7.  *Proposed Mission: 100 Climate-Neutral Cities by 2030—by and for the Citizens. Report of the Mission Board for Climate-Neutral and Smart Cities*; European Commission: Brussels, Belgium, 2020.

8.  Monteiro, C.S.; Causone, F.; Cunha, S.; Pina, A.; Erba, S. Addressing the challenges of public housing retrofits. *Energy Procedia* **2017**, *134*, 442–451. [CrossRef]

9.  Calzada, I. Replicating Smart Cities: The City-to-City Learning Programme in the Replicate EC-H2020-SCC Project. *Smart Cities* **2020**, *3*, 978–1003. [CrossRef]

10. *Europe towards Positive Energy Districs. A Compilation of Projects towards Sustainable Urbanization and the Energy Transition*; Urban Europe: Vienna, Austria, 2020.

11. Amending Directive 2010/31/EU on the Energy Performance of Buildings and Directive 2012/27/EU on Energy Efficiency. In Proceedings of the Directive (EU) 2018/844 of the European Parliament and of the Council, Brussels, Belgium, 30 May 2018.

12. *Horizon Work Programe 2018–2020. Secure, Clean and Efficient Energy*; European Commission: Brussels, Belgium, 2020.

13. *Framework Definition for Positive Energy Districts and Neighbourhoods*; Urban Europe: Vienna, Austria, 2019.

14. Lindholm, O.; Rehman, H.U.; Reda, F. Positioning Positive Energy Districts in European Cities. *Buildings* **2021**, *11*, 19. [CrossRef]

15. Alpagut, B.; Akyürek, Ö.; Mitre, E.M. Positive Energy Districts Methodology and Its Replication Potential. *Proceedings* **2019**, *20*, 8. [CrossRef]

16. Moreno, A.G.; Vélez, F.; Alpagut, B.; Hernández, P.; Montalvillo, C.S. How to Achieve Positive Energy Districts for Sustainable Cities: A Proposed Calculation Methodology. *Sustainability* **2021**, *13*, 710. [CrossRef]

17. Good, N.; Ceseña, E.M.; Mancarella, P. Energy Positivity and Flexibility in Districts. In *Energy Positive Neighborhoods and Smart Energy Districts*; Elsevier BV: Amsterdam, The Netherlands, 2017; pp. 7–30.

18. Skaar, C.; Labonnote, N.; Gradeci, K. From Zero Emission Buildings (ZEB) to Zero Emission Neighbourhoods (ZEN): A Mapping Review of Algorithm-Based LCA. *Sustainability* **2018**, *10*, 2405. [CrossRef]

19. *Zero Emission Neighbourhoods in Smart Cities. Definition, Key Performance Indicators and Assessment Criteria*; SINTEF and NTNU: Trondheim, Norway, 2018.

20. IPCC. *Global Warming of 1.5 °C. An IPCC Special Report on the Impacts of Global Warming of 1.5 °C above Pre-Industrial Levels and Related Global Greenhouse Gas Emission Pathways, in the Context of Strengthening the Global Response to the Threat of Climate Change, Sustainable Development, and Efforts to Eradicate Poverty*; Masson-Delmotte, V.P., Zhai, H.-O., Pörtner, D., Roberts, J., Skea, P.R., Shukla, A., Pirani, W., Moufouma-Okia, C., Péan, R., Pidcock, S., et al., Eds.; SINTEF and NTNU: Trondheim, Norway, 2018.

21. World Urbanization Prospects: The 2018 Revision. Available online: https://population.un.org/wup/Publications/Files/WUP2018-KeyFacts.pdf (accessed on 3 July 2021).

22. Marquard, E.; Bartke, S.; Font, J.G.I.; Humer, A.; Jonkman, A.; Jürgenson, E.; Marot, N.; Poelmans, L.; Repe, B.; Rybski, R.; et al. Land Consumption and Land Take: Enhancing Conceptual Clarity for Evaluating Spatial Governance in the EU Context. *Sustainability* **2020**, *12*, 8269. [CrossRef]

23. European Commission. Directorate General for the Environment, University of the West of England (UWE). Science Communication Unit. No Net Land Take by 2050? Available online: http://ec.europa.eu/science-environment-policy (accessed on 31 July 2021).

24. International Energy Agency. Achieving Net-Zero Emissions by World Energy Outlook Report Extract. Available online: https://www.iea.org/reports/world-energy-outlook-2020/achieving-net-zero-emissions-by-2050 (accessed on 3 July 2021).

25. Sahakian, M.; Wilhite, H. Making practice theory practicable: Towards more sustainable forms of consumption. *J. Consum. Cult.* **2013**, *14*, 25–44. [CrossRef]

26. Axon, S. "Keeping the ball rolling": Addressing the enablers of, and barriers to, sustainable lifestyles. *J. Environ. Psychol.* **2017**, *52*, 11–25. [CrossRef]

27. Amending Di-rectives 2009/125/EC and 2010/30/EU and Repealing Directives 2004/8/EC and 2006/32/EC. In Proceedings of the Directive 2012/27/EU of the European Parliament and of the Council on Energy Efficiency, Brussels, Belgium, 25 October 2012.

28. *Directive 2010/31/EU of the European Parliament and of the Council*; Publications Office of the European Union: Luxembourg, 2010. Available online: https://eur-lex.europa.eu/legal-content/EN/TXT/PDF/?uri=CELEX:32010L0031&from=IT (accessed on 30 July 2021).

29. Sartori, I.; Napolitano, A.; Voss, K. Net zero energy buildings: A consistent definition framework. *Energy Build.* **2012**, *48*, 220–232. [CrossRef]

30. Erba, S.; Pagliano, L.; Shandiz, S.C.; Pietrobon, M. Energy consumption, thermal comfort and load match: Study of a monitored nearly Zero Energy Building in Mediterranean climate. *IOP Conf. Ser. Mater. Sci. Eng.* **2019**, *609*, 062026. [CrossRef]

31. Pagliano, L.; Erba, S.; Peuportier, B. Definition of Indicators and Assessment Methods for Cost Effective nZEB and Energy+ Buildings—An AZEB Projects Report. 2019. Available online: https://cordis.europa.eu/project/id/754174 (accessed on 30 July 2021).

32. Hermelink, A.; Pagliano, L.; Voss, K.; Zangheri, P.; Schimschar, S.; Armani, R.; Voss, K.; Musall, E. *Towards Nearly Zero-Energy Buildings-Definition of Common Principles under the EPBD—Final Report*; European Commission: Brussels, Belgium, 2013.

33. Erhorn-Kluttig, H.; Erhorn, H. National applications of the NZEB Definition–The Complete Overview. In *Concerted Action Energy Performance of Buildings*; Fraunhofer Institute for Building Physics: Stuttgart, Germany, 2018.

34. Attia, S.; Eleftheriou, P.; Xeni, F.; Morlot, R.; Ménézo, C.; Kostopoulos, V.; Betsi, M.; Kalaitzoglou, I.; Pagliano, L.; Cellura, M.; et al. Overview and future challenges of nearly zero energy buildings (nZEB) design in Southern Europe. *Energy Build.* **2017**, *155*, 439–458. [CrossRef]

35. EU HProject: AZEB—Affordable Zero Energy Buildings. Available online: https://azeb.eu/ (accessed on 3 July 2021).

36. *ISO 52000-1:Energy Performance of Buildings–Overarching EPB Assessment–Part 1: General Framework and Procedures*; International Organization for Standardization: Geneva, Switzerland, 2017.

37. Shnapp, S.; Paci, D.; Bertoldi, P. *Enabling Positive Energy Districts across Europe: Energy Efficiency Couples Renewable Energy*; LU: European Commission. Joint Research Centre: Brussels, Belgium, 2020.

38. Le Quartier de la Fleuriaye à Carquefou. Available online: http://www.quartierlafleuriaye.fr/ (accessed on 3 July 2021).

39. Fornebu, Bærum–ZEN Pilot Project. Available online: https://fmezen.no/fornebu-baerum/ (accessed on 3 July 2021).

40. NyBy–Ny Flyplass (New City–New Airport), Bodø–ZEN Pilot Project. Available online: https://fmezen.no/airport-redevelopment-bodo/ (accessed on 3 July 2021).

41. Agliardi, E.; Cattani, E.; Ferrante, A. Deep energy renovation strategies: A real option approach for add-ons in a social housing case study. *Energy Build.* **2018**, *161*, 1–9. [CrossRef]

42. Semprini, G.; Gulli, R.; Ferrante, A. Deep regeneration vs shallow renovation to achieve nearly Zero Energy in existing buildings. *Energy Build.* **2017**, *156*, 327–342. [CrossRef]

43. Shnapp, S.; Sitjà, R.; Laustsen, J. *What is a Deep Renovation Definition?* Global Buildings Performance Network (GBPN): Paris, France, 2013.

44. European Parliament. Report on the Proposal for a Directive of the European Parliament and of the Council on Energy Efficiency and Repealing Directives 2004/8/EC and 2006/32/EC. Available online: https://www.europarl.europa.eu/doceo/document/A-7-2012-0265_EN.html2012 (accessed on 30 July 2021).

45. Galvin, R. Making the 'rebound effect' more useful for performance evaluation of thermal retrofits of existing homes: Defining the 'energy savings deficit' and the 'energy performance gap'. *Energy Build.* **2014**, *69*, 515–524. [CrossRef]

46. Zou, P.X.; Xu, X.; Sanjayan, J.; Wang, J. Review of 10 years research on building energy performance gap: Life-cycle and stakeholder perspectives. *Energy Build.* **2018**, *178*, 165–181. [CrossRef]

47. International Performance Measurement and Verification Protocol. Available from: Concepts and Options for Determining Energy and Water Savings, Vol. 1, International Performance Measurement and Verification Protocol Committee. Available online: http://www.evo-world.org (accessed on 3 July 2021).

48. Noris, F.; Napolitano, A.; Lollini, R. *Measurement and Verification Protocol for Net Zero Energy Buildings*; IEA SHC/ECBS Task 40/Annex 52; International Energy Agency Solar Heating and Cooling Program; EURAC Research: Bolzano, Italy, 2013.

49. Mavrigiannaki, A.; Gobakis, K.; Kolokotsa, D.; Kalaitzakis, K.; Pisello, A.; Piselli, C.; Gupta, R.; Gregg, M.; Laskari, M.; Saliari, M.; et al. Measurement and Verification of Zero Energy Settlements: Lessons Learned from Four Pilot Cases in Europe. *Sustainability* **2020**, *12*, 9783. [CrossRef]

50. Salvia, G.; Morello, E.; Rotondo, F.; Sangalli, A.; Causone, F.; Erba, S.; Pagliano, L. Performance Gap and Occupant Behavior in Building Retrofit: Focus on Dynamics of Change and Continuity in the Practice of Indoor Heating. *Sustainability* **2020**, *12*, 5820. [CrossRef]

51. Cozza, S.; Chambers, J. *GAPxPLORE: Energy Performance Gap in Existing, New, and Renovated Buildings. Learning from Large-Scale Datasets*; Office Fédéral de l'Énergie OFEN: Geneva, Switzerland, 2019.

52. Bulc, V. Speech at Conference on "Multimodal Sustainable Transport: Which Role for the Internalisation of External Costs?". Available online: https://www.linkedin.com/pulse/speech-conference-multimodal-sustainable-transport-which-violeta-bulc/ (accessed on 3 July 2021).

53. Krogstrup, S.; Oman, W. *Macroeconomic and Financial Policies for Climate Change Mitigation: A Review of the Literature*; International Monetary Fund: Bretton Woods, NH, USA, 2019.

54. *EN ISO 52016-1:Energy Performance of Buildings—Energy needs for Heating and Cooling, Internal Temperatures and Sensible and Latent Heat Loads-Part 1: Calculation Procedures*; European Committee for Standardization: Brussels, Belgium, 2017.

55. Huber, M.; Dimkova, D.; Hamacher, T. Integration of wind and solar power in Europe: Assessment of flexibility requirements. *Energy* **2014**, *69*, 236–246. [CrossRef]

56. Jensen, S.Ø.; Marszal, A.J.; Lollini, R.; Pasut, W.; Knotzer, A.; Engelmann, P.; Stafford, A.; Reynders, G. IEA EBC Annex 67 Energy Flexible Buildings. *Energy Build.* **2017**, *155*, 25–34. [CrossRef]

57. Hedman, Å.; Rehman, H.; Gabaldón, A.; Bisello, A.; Albert-Seifried, V.; Zhang, X.; Guarino, F.; Grynning, S.; Eicker, U.; Neumann, H.-M.; et al. IEA EBC Annex83 Positive Energy Districts. *Buildings* **2021**, *11*, 130. [CrossRef]

58. Junker, R.G.; Azar, A.G.; Lopes, R.A.; Lindberg, K.B.; Reynders, G.; Relan, R.; Madsen, H. Characterizing the energy flexibility of buildings and districts. *Appl. Energy* **2018**, *225*, 175–182. [CrossRef]

59. Vigna, I.; Pernetti, R.; Pasut, W.; Lollini, R. New domain for promoting energy efficiency: Energy Flexible Building Cluster. *Sustain. Cities Soc.* **2018**, *38*, 526–533. [CrossRef]

60. Reynders, G.; Lopes, R.A.; Marszal, A.J.; Aelenei, D.; Martins, J.; Saelens, D. Energy flexible buildings: An evaluation of definitions and quantification methodologies applied to thermal storage. *Energy Build.* **2018**, *166*, 372–390. [CrossRef]

61. Foteinaki, K.; Li, R.; Heller, A.; Rode, C. Heating system energy flexibility of low-energy residential buildings. *Energy Build.* **2018**, *180*, 95–108. [CrossRef]

62. Favre, B.; Peuportier, B. Application of dynamic programming to study load shifting in buildings. *Energy Build.* **2014**, *82*, 57–64. [CrossRef]

63. Le Dréau, J.; Heiselberg, P. Energy flexibility of residential buildings using short term heat storage in the thermal mass. *Energy* **2016**, *111*, 991–1002. [CrossRef]

64. Liu, M.; Heiselberg, P. Energy flexibility of a nearly zero-energy building with weather predictive control on a convective building energy system and evaluated with different metrics. *Appl. Energy* **2019**, *233*, 764–775. [CrossRef]

65. Foteinaki, K.; Li, R.; Péan, T.; Rode, C.; Salom, J. Evaluation of energy flexibility of low-energy residential buildings connected to district heating. *Energy Build.* **2020**, *213*, 109804. [CrossRef]

66. Reynders, G.; Diriken, J.; Saelens, D. Generic characterization method for energy flexibility: Applied to structural thermal storage in residential buildings. *Appl. Energy* **2017**, *198*, 192–202. [CrossRef]

67. Johra, H.; Heiselberg, P.; Le Dréau, J. Influence of envelope, structural thermal mass and indoor content on the building heating energy flexibility. *Energy Build.* **2019**, *183*, 325–339. [CrossRef]

68. Reynders, G.; Nuytten, T.; Saelens, D. Potential of structural thermal mass for demand-side management in dwellings. *Build. Environ.* **2013**, *64*, 187–199. [CrossRef]

69. Reilly, A.; Kinnane, O. The impact of thermal mass on building energy consumption. *Appl. Energy* **2017**, *198*, 108–121. [CrossRef]

70. Stinner, S.; Huchtemann, K.; Müller, D. Quantifying the operational flexibility of building energy systems with thermal energy storages. *Appl. Energy* **2016**, *181*, 140–154. [CrossRef]

71. Kuczyński, T.; Staszczuk, A. Experimental study of the influence of thermal mass on thermal comfort and cooling energy demand in residential buildings. *Energy* **2020**, *195*, 116984. [CrossRef]

72. Orosa, J.A.; Oliveira, A. A field study on building inertia and its effects on indoor thermal environment. *Renew. Energy* **2012**, *37*, 89–96. [CrossRef]

73. Panão, M.O.; Mateus, N.M.; Da Graça, G.C. Measured and modeled performance of internal mass as a thermal energy battery for energy flexible residential buildings. *Appl. Energy* **2019**, *239*, 252–267. [CrossRef]

74. Kensby, J.; Trüschel, A.; Dalenbäck, J.-O. Potential of residential buildings as thermal energy storage in district heating systems—Results from a pilot test. *Appl. Energy* **2015**, *137*, 773–781. [CrossRef]

75. *EN 16798-1:Energy Performance of Buildings-Ventilation for Buildings-Part 1: Indoor Environmental Input Parameters for Design and Assessment of Energy Performance of Buildings Addressing Indoor Air Quality, Thermal Environment, Lighting and Acoustics-Module M1-6*; European Committee for Standardization: Brussels, Belgium, 2019.

76. *ANSI/ASHRAE Standard 55-Thermal Environmental Conditions for Human Occupancy*; American Society of Heating, Refrigerating and Air-Conditioning Engineers: Atlanta, GA, USA, 2020.

77. Illich, I. *Energy and Equity*; Harper & Row: New York, NY, USA, 1978.

78. Goldemberg, J.; Johansson, T.B.; Reddy, A.K.N.; Williams, R.H. Basic Needs and Much More with One Kilowatt per Capita. *Ambio* **1985**, *14*, 190–200.

79. Sachs, W. *Die vier E's-Merkposten für Einen Maß-Vollen Wirtschaftsstil; in Politische Ökologie*; nr. 33, 69–72; Wuppertal Institut: Wuppertal, Germany, 1993.

80. Sachs, W. *Planet Dialectics: Explorations in Environment and Development*; Zed Books Ltd.: London, UK, 1999.

81. Toulouse, E.; Sahakian, M.; Lorek, S.; Bohnenberger, K.; Bierwirth, A.; Leuser, L. Energy Sufficiency: How Can Research Better Help and Inform Policy-Making? ECEEE Summer Study Proceedings. Available online: https://www.eceee.org/library/conference_proceedings/eceee_Summer_Studies/2019/2-whats-next-in-energy-policy/energy-sufficiency-how-can-research-better-help-and-inform-policy-making/ (accessed on 30 July 2021).

82. Berg, N.J.V.D.; Hof, A.F.; Akenji, L.; Edelenbosch, O.Y.; van Sluisveld, M.A.; Timmer, V.J.; van Vuuren, D.P. Improved modelling of lifestyle changes in Integrated Assessment Models: Cross-disciplinary insights from methodologies and theories. *Energy Strat. Rev.* **2019**, *26*, 100420. [CrossRef]

83. Samadi, S.; Gröne, M.-C.; Schneidewind, U.; Luhmann, H.-J.; Venjakob, J.; Best, B. Sufficiency in energy scenario studies: Taking the potential benefits of lifestyle changes into account. *Technol. Forecast. Soc. Chang.* **2017**, *124*, 126–134. [CrossRef]

84. Creutzig, F.; Roy, J.; Lamb, W.F.; Azevedo, I.M.; De Bruin, W.B.; Dalkmann, H.; Edelenbosch, O.Y.; Geels, F.W.; Grubler, A.; Hepburn, C.; et al. Towards demand-side solutions for mitigating climate change. *Nat. Clim. Chang.* **2018**, *8*, 260–263. [CrossRef]

85. *Plan Directeur de l'Énergie 2020–2030 République et Canton de Genève*; République et Canton de Genève: Genève, Switzerland, 2020. Available online: https://www.ge.ch/document/22488/telecharger (accessed on 31 July 2021).

86. Creutzig, F.; Callaghan, M.W.; Ramakrishnan, A.; Javaid, A.; Niamir, L.; Minx, J.C.; Müller-Hansen, F.; Sovacool, B.K.; Afroz, Z.; Andor, M.; et al. Reviewing the scope and thematic focus of 100 000 publications on energy consumption, services and social aspects of climate change: A big data approach to demand-side mitigation. *Environ. Res. Lett.* **2021**, *16*, 033001. [CrossRef]

87. Raworth, K. *Doughnut Economics. Seven Ways to Think Like a 21st Century Economist*; Joni Praded, Chelsea Green Publishing: White River Junction, VT, USA, 2017.

88. Jackson, T. *Prosperity without Growth. Foundations for the Economy of Tomorrow*, 2nd ed.; Routledge: London, UK, 2016.

89. Burke, M.J. Energy-Sufficiency for a Just Transition: A Systematic Review. *Energies* **2020**, *13*, 2444. [CrossRef]

90. Brown, H.S.; Vergragt, P.J. From consumerism to wellbeing: Toward a cultural transition? *J. Clean. Prod.* **2016**, *132*, 308–317. [CrossRef]

91. Douglas, M.J.; Watkins, S.J.; Gorman, D.R.; Higgins, M. Are cars the new tobacco? *J. Public Health* **2011**, *33*, 160–169. [CrossRef] [PubMed]

92. *Impact de La Pollution de L'air Ambiant Sur la Mortalité en France Métropolitaine. Réduction en Lien Avec le Confinement du Printemps 2020 et Nouvelles Données sur le Poids Total Pour la Période 2016–2019*; Santé Publique France: Paris, France, 2021. Available online: https://www.santepubliquefrance.fr/determinants-de-sante/pollution-et-sante/air/documents/enquetes-etudes/impact-de-pollution-de-l-air-ambiant-sur-la-mortalite-en-france-metropolitaine.-reduction-en-lien-avec-le-confinement-du-printemps-2020-et-nouvelle (accessed on 30 July 2021).

93. Becker, D.A.; Browning, M.H. Chapter 3: Total area greenness is associated with lower per-capita medicare spending, but blue spaces are not. *City Environ. Interact.* **2021**, *2021*, 100063. [CrossRef]

94. Wiedmann, T.; Lenzen, M.; Keyßer, L.T.; Steinberger, J.K. Scientists' warning on affluence. *Nat. Commun.* **2020**, *11*, 1–10. [CrossRef] [PubMed]

95. Ivanova, D.; Barrett, J.; Wiedenhofer, D.; Macura, B.; Callaghan, M.W.; Creutzig, F. Quantifying the potential for climate change mitigation of consumption options. *Environ. Res. Lett.* **2020**, *15*, 093001. [CrossRef]

96. Akenji, L.; Chen, H. *A Framework for Shaping Sustainable Lifestyles. Determinants and Strategies*; UNEP: Nairobi, Kenya, 2016.

97. Fanger, P.O. *Thermal Comfort, Analysis and Applications in Environmental Engineering*; McGraw-Hill Book Company: New York, NY, USA, 1970.

98. de Dear, R. A global database of thermal comfort field experiments. *ASHRAE Trans.* **1998**, *104*, 1141–1152.

99. Ličina, V.F.; Cheung, T.; Zhang, H.; de Dear, R.; Parkinson, T.; Arens, E.; Chun, C.; Schiavon, S.; Luo, M.; Brager, G.; et al. Development of the ASHRAE Global Thermal Comfort Database II. *Build. Environ.* **2018**, *142*, 502–512. [CrossRef]

100. Fanger, P.O.; Toftum, J. Extension of the PMV model to non-air-conditioned buildings in warm climates. *Energy Build.* **2002**, *34*, 533–536. [CrossRef]

101. Pagliano, L.; Zangheri, P. Comfort models and cooling of buildings in the Mediterranean zone. *Adv. Build. Energy Res.* **2010**, *4*, 167–200. [CrossRef]

102. Carlucci, S.; Erba, S.; Pagliano, L.; de Dear, R. ASHRAE Likelihood of Dissatisfaction: A new right-here and right-now thermal comfort index for assessing the Likelihood of Dissatisfaction according to the ASHRAE adaptive comfort model. *Energy Build.* **2021**, in press. [CrossRef]

103. Lipczynska, A.; Schiavon, S.; Graham, L.T. Thermal comfort and self-reported productivity in an office with ceiling fans in the tropics. *Build. Environ.* **2018**, *135*, 202–212. [CrossRef]

104. He, Y.; Chen, W.; Wang, Z.; Zhang, H. Review of fan-use rates in field studies and their effects on thermal comfort, energy conservation, and human productivity. *Energy Build.* **2019**, *194*, 140–162. [CrossRef]

105. Raftery, P. CBE Fan Tool, Center for the Built Environment, University of California Berkeley. 2019. Available online: Cbe.berkeley.edu/fan-tool (accessed on 30 July 2021).

106. Raftery, P.; Fizer, J.; Chen, W.; He, Y.; Zhang, H.; Arens, E.; Schiavon, S.; Paliaga, G. Ceiling fans: Predicting indoor air speeds based on full scale laboratory measurements. *Build. Environ.* **2019**, *155*, 210–223. [CrossRef]

107. Erba, S.; Sangalli, A.; Pagliano, L. Present and future potential of natural night ventilation in nZEBs. *IOP Conf. Ser. Earth Environ. Sci.* **2019**, *296*. [CrossRef]

108. Dama, A.; De Lena, E.; Masera, G.; Pagliano, L.; Ruta, M.; Zangheri, P. Design and Passive Strategies Optimization Towards Zero Energy Target: The Case Study of an Experimental Office Building in Milan. *Simul. Optim. Conf.* **2014**, *7*. [CrossRef]

109. Sfakianaki, A.; Santamouris, M.; Hutchins, M.; Nichol, F.; Wilson, M.; Pagliano, L.; Pohl, W.; Alexandre, J.; Freire, A. Energy Consumption Variation due to Different Thermal Comfort Categorization Introduced by European Standard EN 15251 for New Building Design and Major Rehabilitations. *Int. J. Vent.* **2011**, *10*, 195–204. [CrossRef]

110. Arens, E.; Humphreys, M.A.; de Dear, R.; Zhang, H. Are 'class A' temperature requirements realistic or desirable? *Build. Environ.* **2010**, *45*, 4–10. [CrossRef]

111. Alfano, F.R.D.; Palella, B.I.; Riccio, G. The role of measurement accuracy on the thermal environment assessment by means of PMV index. *Build. Environ.* **2011**, *46*, 1361–1369. [CrossRef]

112. International Standard Organization. *ISO 7730:Ergonomics of the Thermal Environment—Analytical Determination and Interpretation of Thermal Comfort Using Calculation of the PMV and PPD Indices and Local Thermal Comfort Criteria*; ISO: Geneva, Switzerland, 2005.

113. Aliagha, G.U.; Cin, N.Y. Perceptions of Malaysian Office Workers on the Adoption of the Japanese Cool Biz Concept of Energy Conservation. *J. Asian Afr. Stud.* **2013**, *48*, 427–446. [CrossRef]

114. The District of Florés Malacca. Available online: https://www.smartweb.re/envirobat/files/fiches_envirobat_reunion/logements/FICHE_ENVIROBAT_Reunion_FLORES_MALACCA.pdf (accessed on 3 July 2021).

115. Kleerekoper, L.; Esch, M.P.-V.; Salcedo, T.B. How to make a city climate-proof, addressing the urban heat island effect. *Resour. Conserv. Recycl.* **2012**, *64*, 30–38. [CrossRef]

116. Mariotti, I.; Akhavan, M.; Rossi, F. The preferred location of coworking spaces in Italy: An empirical investigation in urban and peripheral areas. *Eur. Plan. Stud.* **2021**, 1–23. [CrossRef]

117. Akhavan, M.; Mariotti, I.; Astolfi, L.; Canevari, A. Coworking Spaces and New Social Relations: A Focus on the Social Streets in Italy. *Urban Sci.* **2018**, *3*, 2. [CrossRef]

118. Guth, C. The New Cycle Network and Speed Limit Areas in Paris. First Results and Lessons Learnt. Available online: https://www.ridef2.com/uploads/9/5/1/2/95124774/20191216_ms_guth_on_plan_velo_paris_x_ridef_milano.pdf (accessed on 3 July 2021).
119. Kraus, S.; Koch, N. Provisional COVID-19 infrastructure induces large, rapid increases in cycling. *Proc. Natl. Acad. Sci. USA* **2021**, *118*. [CrossRef]
120. Jackson, T. Zero Carbon Sooner—The Case for an Early Zero Carbon Target for the UK. Available online: https://www.cusp.ac.uk/wp-content/uploads/WP18%E2%80%94Zero-carbon-sooner.pdf (accessed on 30 July 2021).
121. Mundaca, L.; Ürge-Vorsatz, D.; Wilson, C. Demand-side approaches for limiting global warming to 1.5 °C. *Energy Effic.* **2019**, *12*, 343–362. [CrossRef]
122. Santamouris, M.; Kolokotsa, D. *Urban Climate Mitigation Techniques*; Routledge: London, UK, 2016.
123. Ip, K.; She, K. Waste heat recovery from showers: Case study of a university sport facility in the UK. In Proceedings of the Water Efficiency Conference, Coventry UK, 7–9 September 2016.
124. *EN 14825:Air Conditioners, Liquid Chilling Packages and Heat Pumps, with Electrically Driven Compressors, for Space Heating and Cooling-Testing and Rating at Part Load Conditions and Calculation of Seasonal Performance*; European Committee for Standardization: Brussels, Belgium, 2018.
125. Causone, F.; Carlucci, S.; Ferrando, M.; Marchenko, A.; Erba, S. A data-driven procedure to model occupancy and occupant-related electric load profiles in residential buildings for energy simulation. *Energy Build.* **2019**, *202*, 109342. [CrossRef]
126. International Organization for Standardization. *ISO 15927-6:Hygrothermal Performance of Buildings—Calculation and Presentation of Climatic Data Accumulated Temperature Differences (Degree-Days)*; International Organization for Standardisationn (IOS): Geneva, Switzerland, 2007.
127. Pagliano, L.; Armani, R.; Erba, S.; Sangalli, A. Highly Insulated Buildings as a Crucial Element for Smart Cities, Grid Balancing and Energy Storage for Renewables. Knauf Insulation Report. 2020. Available online: https://www.knaufinsulation.com/news/thermal-insulation-of-buildings-for-a-smart-and-renewable-energy-system (accessed on 31 July 2021).
128. Joint Research Centre. *Institute for Energy and Transport. Individual Mobility: From Conventional to Electric Cars*; LU: Publications Office, European Commission: Brussels, Belgium, 2015.
129. Festa, M. Rapporto Immobiliare Il Settore Residenziale. 2019. Available online: https://www.agenziaentrate.gov.it/portale/documents/20143/263076/rapporto+immobiliare2019+ri_RI2019_Residenziale_20190523.pdf/a175f856-2363-dda7-da64-b1c0e544eb12 (accessed on 31 July 2021).
130. Lenoir, A.; Thellier, F.; Garde, F. Towards Net Zero Energy Buildings in Hot Climate, Part 2: Experimental Feedback. *ASHRAE Trans.* **2011**, *117*, 8.
131. Relazione Annuale. Stato dei Servizi 2019. ARERA 2019. Available online: https://www.arera.it/allegati/relaz_ann/19/RA19_volume1.pdf (accessed on 30 July 2021).
132. Brischke, L.-A.; Lehmann, F.; Leuser, L.; Thomas, S.; Baedeker, C. Energy Sufficiency in Private Households Enabled by Adequate Appliances, ECEEE Summer Study Proceedings. 2015. Available online: https://epub.wupperinst.org/frontdoor/deliver/index/docId/5932/file/5932_Brischke.pdf (accessed on 30 July 2021).
133. Consigli Piani di Cottura. Available online: https://www.topten.ch/private/adviser/consigli-piani-di-cottura (accessed on 3 July 2021).
134. European Commission. Ecodesign and Energy Labelling Preparatory Study on Electric Kettles. Available online: https://ec.europa.eu/energy/studies_main/preparatory-studies/ecodesign-and-energy-labelling-preparatory-study-electric-kettles_en (accessed on 3 July 2021).
135. Patrão, C.; Rivet, L.; Fong, J.; de Almeida, A. Energy Efficient Elevators and Escalators. ECEEE 2009 Summer Study Act! Innovate! Deliver! Reducing Energy Demand Sustainably Proceedings. 2009. Available online: https://www.eceee.org/static/media/uploads/site-2/library/conference_proceedings/eceee_Summer_Studies/2009/Panel_4/4.037/paper.pdf (accessed on 30 July 2021).
136. SIA 2024:2015 Costruzione. Dati D'utilizzo di Locali per L'energia e L'impiantistica Degli Edifici. 2015. Available online: http://shop.sia.ch/normenwerk/architekt/sia%202024/i/D/Product (accessed on 31 July 2021).
137. *Abitudini di Guida Degli Italiani nel 2018 in Seguito all'analisi dei Dati Delle Scatole Nere Installate Nelle Automobile*; Osservatorio UnipolSai Assicurazioni: Bologna, Italy, 2019.
138. *L'Efficienza Energetica nei Trasporti*; ENEA; ENEA (Agenzia Nazionale per le Nuove Tecnologie, l'Energia e lo Sviluppo Economico Sostenibile): Roma, Italy, 2011.
139. Šúri, M.; Huld, T.; Dunlop, E.D.; Ossenbrink, H.A. Potential of solar electricity generation in the European Union member states and candidate countries. *Sol. Energy* **2007**, *81*, 1295–1305. [CrossRef]

*Article*

# A GIS-Based Multicriteria Assessment for Identification of Positive Energy Districts Boundary in Cities

Beril Alpagut [1,*], Arantza Lopez Romo [2], Patxi Hernández [2], Oya Tabanoğlu [1] and Nekane Hermoso Martinez [2]

[1]  Demir Energy, Smart and Sustainable Cities Department, Istanbul 34718, Turkey; otabanoglu@demirenerji.com
[2]  Tecnalia, Basque Research and Technology Alliance (BRTA), 48160 Derio, Spain; arantza.lopez@tecnalia.com (A.L.R.); patxi.hernandez@tecnalia.com (P.H.); nekane.hermoso@tecnalia.com (N.H.M.)
*   Correspondence: balpagut@demirenerji.com; Tel.: +90-533-237-6200

**Abstract:** Discussions regarding the definition of Positive Energy Districts and the concept of a boundary are still being actively held. Even though there are certain initiatives working on the boundary limitations for PEDs, there is no methodology or tool developed for selecting peculiar spaces for future PED implementations. The paper focuses on a flexible GIS-based Multicriteria assessment method that identifies the most suitable areas to reach an annual positive non-renewable energy balance. For that purpose, a GIS-based tool is developed to indicate the boundary from an energy perspective harmonized with urban design and land-use planning. The method emphasizes evaluation through economic, social, political, legal, environmental, and technical criteria, and the results present the suitability of areas at macro and micro scales. The current study outlines macro-scale analyses in six European cities that represent Follower Cities under the MAKING-CITY H2020 project. Further research will be conducted for micro-scale analyses and the outcomes will pursue a technology selection process.

**Keywords:** positive energy districts; PED boundary; multi-criteria decision analyses; geographic information systems; GIS overlay analyses

**Citation:** Alpagut, B.; Lopez Romo, A.; Hernández, P.; Tabanoğlu, O.; Hermoso Martinez, N. A GIS-Based Multicriteria Assessment for Identification of Positive Energy Districts Boundary in Cities. *Energies* **2021**, *14*, 7517. https://doi.org/10.3390/en14227517

Academic Editor: Paola Clerici Maestosi

Received: 3 August 2021
Accepted: 18 September 2021
Published: 11 November 2021

**Publisher's Note:** MDPI stays neutral with regard to jurisdictional claims in published maps and institutional affiliations.



## 1. Introduction

The potential for high energy savings in building stock has long been recognized, and different policies have been established for achieving savings at the European Level. The original Energy Performance of Buildings Directive (EPBD) in 2002 introduced more strict building regulations and energy certification, while the policies have evolved towards the requirement of "nearly zero energy building", in the EPBD 2010 update [1], where it is defined as *"a building that has a very high energy performance. The nearly zero or very low amount of energy required should be covered to a very significant extent by energy from renewable sources, including energy from renewable sources produced on-site or nearby"*. This directive already required that the calculation of energy performance in buildings should consider district heating or cooling solutions, as these solutions can have potential benefits for the performance of buildings. The amendment of the EPBD Directive in 2018 included no additional requirements to evaluate the district-level energy performance, although it does state that the Commission should review the EPBD before January 2026, to *"examine in what manner Member States could apply integrated district or neighborhood approaches in Union building and energy efficiency policy"*.

In the meantime, there are different EU initiatives promoting strategies and technologies for improving energy efficiency and increasing renewable energy at the district level. For example, the Strategic Energy Technology (SET) Plan Action 3.2 [2], *"Energy Districts and Neighborhoods for Sustainable Urban Development"* aims to support the planning, deployment, and replication of 100 Positive Energy Neighborhoods by 2025. In order to reach this target, the concepts of Positive Energy Blocks (PEBs) and Positive Energy Districts (PEDs) have initially emerged from the EU Horizon 2020 Smart Cities and Communities project calls [3], and a number of research projects such as MAKING-CITY [4], POcityF [5], ATELIER [6], CityxChange [7], SPARCs [8], and RESPONSE [9] are being funded to test and realize Positive Energy Districts (PEDs) in "Lighthouse" cities, and support "follower cities" to replicate the experience. This paper addresses a methodology that has been developed within the MAKING-CITY project, in order to provide knowledge and experience to the participant cities to identify and select PEDs. The selection of suitable areas is the first step towards planning and realization of successful PEDs, which aims to help cities in their efforts to integrate energy planning within urban design and planning. It is expected that the proposed method could serve as an example and could be replicable in different EU city contexts. While the above-mentioned projects work on PED design and development, PED districts are generally individually selected taking into account different parameters, but no structured process for the identification and selection of areas within the cities with the potential to become energy positive has been defined. The proposed method in this paper relates to the various approaches that are mentioned in the literature for identifying suitable areas for energy planning, benefitting from GIS-based methodologies [10–15]. The selection of the areas has to be adapted to the city's characteristics, considering both spatial/physical and technical characteristics of the cities, together with socio-economic, environmental, legal, and regulatory issues. Following this approach, a flexible GIS-based Multicriteria assessment method is developed that identifies the most suitable areas to reach an annual positive non-renewable energy balance. For that purpose, a GIS-based tool is generated to indicate the boundary from an energy perspective harmonized with urban design and land-use planning. The method emphasizes evaluation through economic, social, political, legal, environmental, and technical criteria and results present the suitability of areas at macro and micro scales.

Moreover, the definition of the PED concept boundary is still at the research stage and scenarios regarding the energy delivered, energy use and demand, and RES on-site still vary from project to project. Gabaldon et al. state that the key concept of PED is that of a district that produces more energy from RES than what is needed to fulfil the district's demand, being able to export this energy surplus to another part of the city [16]. According to the MAKING-CITY project, a Positive Energy District (PED) is *"an urban area with clear boundaries, consisting of buildings of different typologies that actively manage the energy flow between them and the larger energy system to reach an annual positive non-renewable energy balance"* [4]. Aligned with this definition, a methodology for PED design [17] is developed, and parameters concerning resources, energy planning, and energy infrastructure, so-called PED Analytical Components, are collected within city boundaries, evaluated through a multi-criteria analysis, and GIS-based analyses at macro and micro scales are conducted to identify peculiar PED boundaries in cities. In this publication, macro scale analysis has been examined, detailing the type of data and characteristics that need to be gathered at the city level to allow for the selection of areas most suitable for PED development. A further step in the methodology should deal with a more detailed micro-scale analysis, where more detailed data with more resolution for each specific building or group of buildings (e.g., energy use and energy systems, socioeconomic characteristics) would be gathered and processed. This additional micro-scale analysis would support preliminary PED design and viability studies, but will not be covered in this paper, which will focus only on the macro-scale analysis for identification of PED boundaries.

## 2. Methodology for Identifying PED Concept Boundary in Cities

If a wide range of stakeholders are active in making a decision (when it comes to PED arguments), utilizing a Multi Criteria Decision Analysis has multiple benefits as it allows for considering a diverse set of values, targets, and interests from the various actors involved in planning, designing, and implementing PEDs. Integrating spatial analysis with MCDA is impactful in terms of evaluating multiple criteria for defining different scenarios on geographic data models [18]. Several GIS-MCDA applications refer to site selection, scenario evaluation, land suitability, impact assessment, and location allocation to a variety of sectors [19–21].

The methodology proposed in this paper for defining PED concept boundaries consists of three phases that synthesize the spatial information system of the cities together with technical, economic, social, environmental, political, and legal frameworks within their context. The phases of the methodology are summarized in Figure 1.

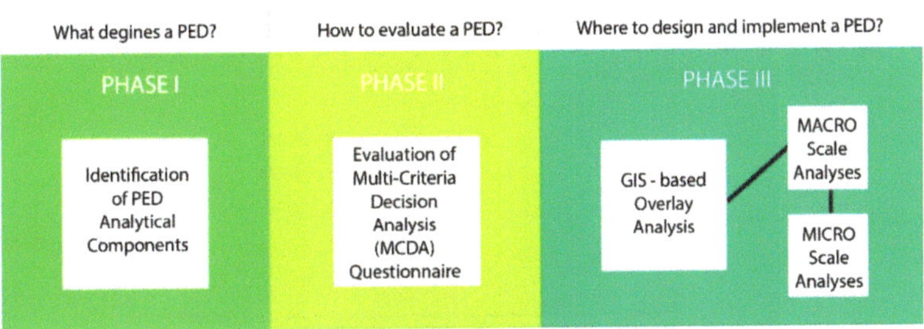


**Figure 1.** Phases of methodology for identifying PED concept boundary in cities.

### 2.1. Phase I: Identification of PED Analytical Components (Critical Elements)

The First Phase is the identification of PED Analytical components that play a key role in the selection of peculiar and efficient PED concept boundary in cities. The challenge is that local energy production and distribution, connected with digitalization, have not previously been a part of the integrated urban planning and design approaches, while they have included many other environmental and social topics [22]. The proposed methodology for identifying PED concept boundaries in cities underlines energy sustainability in urban planning, land-use planning, and urban design, and therefore requires deep analysis at the city level, down to the neighborhood, district, and building levels. Harmonization of these diverse modes of spatial planning with energy planning is the main aspect of this methodology. Likewise, this methodology indicates that inclusiveness, co-creation, and participatory planning shall rule the energy transition. Inclusive cities are powerful by avoiding marginalization, which compromises the richness of interaction upon which cities depend [23].

The proposed methodology therefore demands extensive analyses on resources, urban planning, land-use planning, energy planning in physical and virtual infrastructure, and socio-economic and socio-cultural aspects. All the abovementioned aspects have been classified in the following six categories [22]:

1. Resource analysis.
2. Urban macro-form analysis.
3. Land-use context.
4. Energy infrastructure analysis.
5. Energy services.
6. Social structure.

Data regarding these six categories need to be collected in GIS-based spatial data, since the analyses that are going to be conducted will utilize overlay analyses within GIS-based software.

The first category is the resource availability in cities, consisting of solar, wind, earth, geothermal water, surface water, biomass, and waste heat potential. The resources, their spatial references, and detailed descriptions are displayed in Table 1.

**Table 1.** Category 1: Resource availability.

|   | **Sub-Categories** | **Spatial Reference** |
|---|---|---|
| A | Solar efficient zones | Existing solar energy investment zones<br>Potential solar energy investment zones |
| B | Wind efficient zones | Existing wind energy investment zones<br>Potential wind energy investment zones |
| C | Extensive ground coupling potential for cooling and heating purposes | Geomorphological structure<br>Soil types, formations, ground maps of the city |
| D | Geothermal Water | Geothermal water impact area<br>Potential Geothermal Investment zones |
| E | Water resources | Potential surface water resources for hydropower generation<br>Potential water resources utilized for heat source as heating/cooling purposes<br>Water surfaces with evaporative potential |
| F | Biomass | Potential energy generation areas by biomass |
| G | Waste heat potential | Waste heat energy generation potential |

The second category is related to the urban planning and strategies of the city, so-called "urban macroform". This category gives detailed information derived from city plans and strategies. New development, retrofitting, infill, and re-use/transformation areas are sub-categories to be identified under this category and are described in Table 2.

**Table 2.** Category 2: Urban macroform.

|   | **Sub-Categories** | **Spatial Reference** |
|---|---|---|
| A | New Development Areas | New Development Zones |
| B | Retrofitting Areas | Old Building Stock areas |
| C | Infill Areas | Redevelopment or land recycling areas |
| D | Reuse/Transformation Areas | Urban Transformation Areas |

The third category is again related to the urban context of the city by the land-use coverings. Sub-categories are categorized as residential/mixed-use, commercial/industrial/office, active green/open parking lots, public administration, and social/cultural/educational/sport areas that are displayed in Table 3.

**Table 3.** Category 3: Land-usage context.

|   | **Sub-Categories** | **Spatial Reference** |
|---|---|---|
| A | Land Cover in Zonings/Islands | Residential & Mixed-Use Areas<br>Commercial Areas<br>Active Green/Open Parking Lot<br>Public Administration Areas<br>Social/Cultural/Educational/Sport Areas |

The fourth category is the technical/physical infrastructure that analyzes the energy and e-mobility structure of the cities. Off-grid systems may also play an interesting role for selecting PED areas for their potential in energy flexibility and trading opportunities. For this reason, under this category, district heating, the power-heat network, and the e-mobility structure of the cities are considered (Table 4).

**Table 4.** Category 4: Physical infrastructure.

|  | Sub-Categories | Spatial Reference |
|---|---|---|
| A | Heat Grid | Existing District Heating/Cooling Zones |
| B | Power Infrastructure | High/Low voltage power grid and its impact area |
| C | Heat Network | Natural gas pipeline network |
| D | e-Mobility Infrastructure | Existing EV chargers and impact areas |

The fifth category is the potential virtual infrastructure that the cities may test in terms of smart-grid applications. A few cities in Europe have already started testing virtual power plants and their effect on grids. The impact areas of micro-grid applications may have a key role for selecting PED areas (Table 5).

**Table 5.** Category 5: Virtual infrastructure.

|  | Sub-Categories | Spatial Reference |
|---|---|---|
| A | Smart Grid Applications | Impact areas of micro-grids/islands |

The last category is the social structure of the city represented by spatial information. The socio-economic, socio-cultural context of the city is targeted under this category. Human behavior in energy consumption and energy investment is analyzed and their spatial references are identified. More details on the description of the sub-categories may be found in Table 6.

**Table 6.** Category 6: Social structure.

|  | Sub-Categories | Spatial Reference |
|---|---|---|
| A | Current and Projected Population | Population Density identified in Spatial Data<br>Population Projections for new development zones |
| B | Energy Organizations | Impact and organizational areas of energy organizations<br>Self-sufficient districts/neighborhoods |
| C | Communities | Vulnerable Communities/Energy Poverty<br>Cultural Human Behavior |

*2.2. Phase II: Relevant Criteria to Be Considered for Components Evaluation by MCDA Quesitonnaire: City Characteristics and Priorities*

The second phase is the generation of a Multi-Criteria Decision Analyses Questionnaire (MCDA) for supporting cities to evaluate the PED Components from political, economic, social, technical, environmental, and legal points of view.

A detailed MCDA questionnaire was prepared to gather information from the six categories that have been selected in the previous phase. The questionnaire had specific evaluation criteria for each of the sub-categories shown in Tables 1–6, each covering economic, social, technical, environmental, and legal aspects. The MCDA questionnaire aims to guide cities in understanding the relevant spatial references that could define the suitability of an area to become positive and be able to provide an evaluation for each sub-category. The MCDA questionnaire plays an important role in the assessment in five different ways:

1. It standardizes the application of the geoprocessing analysis. All the cities are studied with the same criteria. Different scores are given to each PED Analytical Component depending on the status of the city in each criterion. These characteristics of the questionnaire allow one to both standardize the method and adapt it to the city's needs.
2. It helps to select the PED Analytical Components to be considered in the assessment.
3. It allows for obtaining reclassification scores for each Analytical Component through the consideration of all the criteria that are relevant. This also helps to consolidate the analysis and adapt it accordingly to the city's state of play.

4.  It provides information for prioritization. Therefore, the weighting phase of the methodology is supported by the information provided by the MCDA questionnaire results.
5.  Going beyond the spatial analysis, it helps cities in identifying the strengths and weaknesses of the city in terms of suitability to become positive in the economic, social, technical, environmental, and legal framework. As it is shown in the results, the MCDA questionnaire is relevant for identifying, in a consensual way, issues that are of interest for the city, but the city has not identified yet and/or, in the opposite case, issues not suitable. For example, the legal framework could facilitate or obstruct the implementation of specific solutions that have potential depending on the city's context.

To ensure that cities understand the various concepts and terminology used in the questionnaires, one-to-one discussion sessions were organized with city public officers and technicians. In each session, a specific category was analyzed following the process explained below for the second to seventh steps.

- First: Identify PED analytical components to be considered in the assessment and describe how each of them is going to be analyzed (Section 2.1).
- Second: Identify the criteria relevant for the PED analytical component in the evaluation, in terms of economic, social, technical, environmental, legal, and spatial, in order to assess the PED analytical components in a robust way.
  - For example, in the case of existing solar energy investment zones, economic, technical, and spatial criteria were selected.
- Third: Establish the information that will allow for analyzing each of the selected criteria.
  - Following the example, the size and solar potential of the solar energy investment zone were considered relevant for analyzing the spatial criteria.
- Fourth: Each of the criteria considered were divided in different possibilities and, according to the suitability of these possibilities, scores were given.
  - Following the example, zones with high solar potential, higher than 20,000 m$^2$, were considered the most suitable and, thus, maximum scores were given to this situation. Zones with high solar potential with a size between 10,000 and 20,000 m$^2$ were also considered suitable but with lower scores than the previous situation.
- Fifth: The considered criteria were prioritized. GIS analysis established that the sum of the best possibilities of each criterion was 9. According to this, scores were divided between considered criteria.
  - Following the example, the economic maximum criteria scored 3, technical scored 2, and spatial criteria scored 4a.
- Sixth: The restrictive criteria were identified. This means that if these possibilities are selected, the final score of the raster is determined only with this value.
  - For example, for the spatial criteria of the * Existing solar energy investment zones, if the possibility "There are no zones that fulfil the previous criteria" is selected, meaning that there are no zones owned by the community or the public administration with higher than 10,000 m$^2$ medium-high solar potential, this PED component no longer needs to be considered in the assessment.
- Seventh: Give extra points and/or minus points to some PED components in order to include criteria that were not considered in the previous options.
  - For example, extra points will be given in the case that this possibility exists: "Surrounding the city there are zones owned by the community or the public administration with potential to become solar parks." and minus points will be given in the case that this possibility exists: " The most suitable zones for implementing solar energy are green areas".

    ○    Note that the extra point can nullify the restrictive criteria. For example, if the extra points possibility exists: "Surrounding the city there are zones owned by the community or the public administration with potential to become solar parks", the restrictive criteria " There are no zones that fulfil the previous criteria" related to the spatial availability of zones with medium-high solar potential, is nullified.

All of the PED Analytical Components (categorized under resource availability (RA), Urban Macroform (UM), Land use Context (LU), Technical–Physical infrastructure (TPI), Virtual Infrastructure (VI), and Social Structure (SS) are reviewed by technical experts in the consortium and their advice is taken into consideration while assessing the layers.

A summary of the criteria (economic, EC, social, SO, technical, TC, environmental, EN, legal, LE, spatial, SP) considered for each PED component is given below (Tables 7–12).

**Table 7.** Criteria considered for resource availability PED analytical components analysis.

| RA. Resource Availability | | | | | | |
| --- | --- | --- | --- | --- | --- | --- |
| **PED Analytical Components** | **EC** | **SO** | **TC** | **EN** | **LE** | **SP** |
| Existing Solar Energy Investment Zones | √ | | √ | | | √ |
| Potential Solar Energy Investment Zones | √ | √ | √ | | √ | √ |
| Existing Wind Energy Investment Zones | √ | | √ | | | √ |
| Potential Wind Energy Investment Zones | √ | √ | √ | | | √ |
| Extensive Ground Coupling Potential | √ | | √ | | √ | √ |
| Geothermal Water Impact Area | | √ | √ | | | √ |
| Potential Surface Water Resources for Energy Generation | √ | | √ | | √ | |
| Potential Water Resources for Heating/Cooling Purposes | √ | | √ | | √ | |
| Water Surfaces with Evaporative Potential | | | | | | √ |
| Potential Energy Generation Areas by Biomass | | √ | √ | √ | | √ |
| Waste Heat Potential | | | √ | | | √ |

**Table 8.** Criteria considered for Urban Macroform PED analytical components analysis.

| UM. Urban Macroform | | | | | | |
| --- | --- | --- | --- | --- | --- | --- |
| **PED Analytical Components** | **EC** | **SO** | **TC** | **EN** | **LE** | **SP** |
| New Development Areas | | | √ | | √ | |
| Retrofitting Areas | √ | | | | √ | √ |
| Infill Areas | | | √ | | √ | √ |
| Urban Transformation/Reuse Areas | √ | | | | √ | √ |

**Table 9.** Criteria considered for Land Usage PED analytical components analysis.

| LU. Land Usage Context | | | | | | |
| --- | --- | --- | --- | --- | --- | --- |
| **PED Analytical Components** | **EC** | **SO** | **TC** | **EN** | **LE** | **SP** |
| Residential & Mixed-Use Areas | √ | √ | | | √ | |
| Commercial areas | √ | | | | √ | |
| Active Green/Open Parking Lot | √ | | √ | √ | √ | |
| Public Administration areas | √ | | | | √ | |
| Social/Cultural/Educational/Sport Areas | √ | | | | √ | |

**Table 10.** Criteria considered for Technical–physical infrastructure PED analytical components analysis.

| TPI. Technical—Physical Infrastructure | | | | | | |
|---|---|---|---|---|---|---|
| **PED Analytical Components** | **EC** | **SO** | **TC** | **EN** | **LE** | **SP** |
| Heat Grid | √ | | √ | | √ | √ |
| Power Infrastructure | √ | | √ | | √ | |
| Heat Network | √ | √ | √ | | √ | √ |
| Mobility Infrastructure | √ | | | | √ | |

**Table 11.** Criteria considered for Virtual infrastructure PED analytical components analysis.

| TPI. Technical—Physical Infrastructure | | | | | | |
|---|---|---|---|---|---|---|
| **PED Analytical Components** | **EC** | **SO** | **TC** | **EN** | **LE** | **SP** |
| Smart Grid Applications. Considering Virtual Power Plants, Micro Grid Applications | √ | | √ | | √ | √ |

**Table 12.** Criteria considered for Social PED analytical components analysis.

| LU. Land Usage Context | | | | | | |
|---|---|---|---|---|---|---|
| **PED Analytical Components** | **EC** | **SO** | **TC** | **EN** | **LE** | **SP** |
| Population Density identified in Spatial Data | √ | √ | | | √ | |
| Population Projections for New Development Zones | √ | | | | √ | |
| Impact And Organizational Areas of Energy Organizations | √ | | √ | √ | √ | |
| Self-Sufficient Districts/Neighborhoods or Ecovillages | √ | | | | √ | |
| Cultural Human Behavior | √ | | | | | |
| Vulnerable Communities/Disadvantageous/Urban Poor | √ | | | | √ | |

This multi-criteria, multi-actor evaluation process is conducted as a parallel analysis to the collection and evaluation of GIS-based data. As it was mentioned before, the results from the questionnaire were used for GIS layers' identification, reclassification, and prioritization purposes.

### 2.3. Phase III: GIS-Based Overlay Analysis

Formerly, an overlay analysis (consisting of geoprocessing steps for ARCGIS and QGIS software) is conducted with the help of the results of the MCDA Questionnaire, and the results are displayed at the macro and, in a more detailed approach, micro scales. In the macro-scale analysis, all data are gathered at the zoning/island scale. No building-scale data are integrated in the analysis. A geoprocessing analysis will be conducted in order to realize the spatial weighted overlay and prioritize potential areas to be PEDs. At the end of this phase, two to three potential zones will be further selected to go on with micro-scale analysis in order to define the PED boundary.

Throughout the whole process, the harmonization of urban planning and energy planning is targeted.

Overlay analysis is a group of methodologies applied in optimal site selection or suitability modelling. It is a technique for applying a common scale of values to diverse and dissimilar inputs to create an integrated analysis. Spatial-based overlay analysis often requires the analysis of many different factors [24].

- Geoprocessing Analysis: Collected city data, at the macro scale, need to be organized. Values in each layer are adjusted according to PED Analytical Components criteria. If the layers are in vector format, the layers need to be converted into raster format since overlay analysis and all of its steps work with raster layers that have integer values (Step 1). Afterwards, reclassification of all input layers will be conducted to manage layers in the accepted format (integer values) (Step 2). Finally, all raster layers will be ready to utilize weighted overlay analysis tool in order to conduct a Suitability analysis in the GIS-based process to determine the appropriateness of a PED boundary. Detailed information regarding the steps of geoprocessing analysis may be found in the Advanced GIS Spatial Analysis and Modeling Tools ArcGIS Spatial Analyst [25]. The process is detailed in five steps.

- Step 1. Conversion to rasters Weighted Overlay analyses only allow integer rasters as input, such as a raster of land usage or soil types for the geomorphological structure. Generally, the values of continuous rasters are grouped into ranges that must be allocated with a single value before Step 2.

- Step 2. Raster Reclassification Following the conversion to rasters, the Reclassify tool must be utilized for reclassifying the generated rasters. The reclassification tools reclassify or change cell values to alternative values using a variety of methods. Values or a group of values inside a layer are reclassified according to importance, interest, or similarities as in specified intervals.

As an example, the values inside a layer (regarding one PED analytical component) e.g., Land-usage, could be residential/mixed use/commercial. The importance of land-usage regarding PED implementations depends heavily on the political and economic context. If the city has incentives or legislations for retrofitting the existing residential use in the city, then residential group values would receive the highest value.

According to the Methodology for PED Concept Boundary Identification, the reclassification step is guided by technical expertise from MAKING-CITY partners on specific knowledge and experiences from LHCs' PED designs and implementations. All of the PED Analytical Components (categorized under resource availability (RA), Urban Macroform (UM), Land use Context (LU), Technical and Physical Infrastructure (TPI), Virtual Infrastructure (VI), and Social Structure (SS) are reviewed by technical experts in the consortium and their advice is taken into consideration while assessing the layers.

Regarding technical comments, for each PED analytical component, a buffer zone for the impact area (if the layer is gathered as "point feature" in GIS format) is generated depending on the existing regulations, conducted studies on economic feasibility, promoted subsidies, or incentives available in each city context. The relation between spatial information and technical, political, economic, and social points of view is targeted to achieve more suitable results.

Since the GIS background and infrastructure of cities vary highly from each other (especially on the clarity of data and values indicated inside each layer (matching the PED analytical components), the reclassification methodology aims to be validated in a comprehensive and holistic approach.

- Step 3. Select an evaluation scale The evaluation scale (Table 13) represents the range of suitability; the highest values show one extreme of suitability whereas the lowest values represent the lowest suitability. The input rasters are reclassified as a common measurement scale utilizing the Reclassification tool. An evaluation scale from 1 to 9 is chosen for the current research.

**Table 13.** Evaluation scale variations.

| Evaluation Scale | 1 | 2 | 3 | 4 | 5 | X |
|---|---|---|---|---|---|---|
| 1–3 | Least suitable | Suitable | Most suitable | | | |
| 1–5 | Very low suitability | low suitability | moderate suitability | High suitability | Very high suitability | |
| 1–X | lowest | | | | | highest |

- Step 4. Set scale values The cell values for each input raster in the analysis are assigned values from the evaluation scale. According to the potential of suitability, the default values allocated to each cell may be changed.
- Step 5. Assign weights to input rasters As the last step, after setting scale values, a percentage influence is assigned to each input raster, based on its importance and effect on suitability. The total influence for all rasters must equal 100%. The criteria considered more important than other criteria are weighted, and some criteria may have equal importance in terms of creation potential for PED implementation. Weights of each macro/micro-scale inputs (PED analytical components water resources, geomorphological structure, land use, buildings, etc.) are determined by considering local/regional/national contexts such as laws and regulations, technical and technological aspects, etc., derived from MCDA. Overlapping all Layers: Overlapping all layers refers to Step 3, Step 4, and Step 5 of the geoprocessing process. Regarding the methodology for PED concept boundary identification, the overlay analysis is conducted via setting score values (remap values) for each PED Analytical component with the help of the MCDA Questionnaire completed by the city. As explained in Section 2.2, the MCDA Questionnaire provides the opportunity for cities to understand and evaluate the PED Analytical components from economic, social, environmental, legal, and political points of view. The scores assigned to each layer are transferred to the remap table in the weighted overlay analysis. The influence tables that provides the rasters (all PED Components) can be evaluated and compared with each other by assigning the importance of shares in terms of %, according to the total sum of remapped values and calculated ratios of each layer.
- Step 6. Run the Overlay Tool As a result, the layers are combined and the overlay layer is obtained. Modifying the suitability values or the influence percentages will produce different results for the output suitability raster.

### 3. Case Studies

*3.1. Case Studies Presentation*

The methodology defined to select the areas to reach an annual positive non-renewable energy balance has been tested in six European cities: Bassano del Grappa (Italy), Kadiköy (Turkey), León (Spain), Lublin (Poland), Trenčín (Slovakia), and Vidin (Bulgaria) (Figure 2) [26].

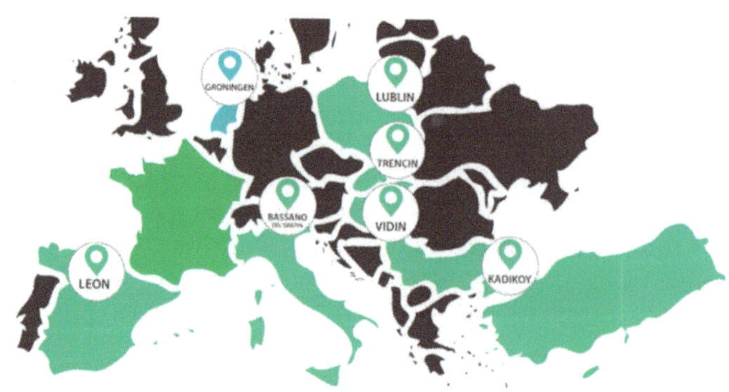


**Figure 2.** MAKING-CITY Project—follower cities.

Cities' main characteristics, primary energy sources currently used, and renewable resources are summarized in Tables 14–16, respectively. The information reflects the European representativeness of the cities selected for the assessment.

**Table 14.** Main characteristics of the cities.

| City | Surface (km$^2$) | Inhabitants | Density (People/km$^2$) | Climate |
|---|---|---|---|---|
| Bassano del Grappa | 46.79 | 43,500 | 929 | Mild |
| Kadıköy | 25.20 | 458,638 | 18,200 | Mediterranean/Black Sea |
| León | 39.2 | 124,772 | 3182 | Oceanic |
| Lublin | 147.4 | 342,039 | 2326 | Oceanic/Continental |
| Trenčín | 82 | 54,916 | 66,979 | European Continental |
| Vidin | 63.22 | 41,583 | 65,777 | Continental |

**Table 15.** Primary energy sources of the cities.

| City | Solid Fossil Fuels (MWh/cap) | Natural Gas (MWh/cap) | Oil and Petroleum (MWh/cap) | Renewables and Biofuels (MWh/cap) | Electricity from the Grid (MWh/cap) |
|---|---|---|---|---|---|
| Bassano del Grappa | 1.52 | 7.35 | 4.22 | 2.29 | 3.70 |
| Kadıköy | 0 | 0.67 | 0 | 0.06 | 2.68 |
| León | 0 | 7.23 | 13.16 | 0.05 | 5.22 |
| Lublin | 0.53 | 3.67 | 0.02 | 4.25 | 3.30 |
| Trenčín | 3.85 | 4.62 | 4.24 | 2.12 | 4.43 |
| Vidin | 1.51 | 0.10 | 1.76 | 1.55 | 6.08 |

**Table 16.** Main characteristics of the renewable sources of the cities.

| City | Average Solar Radiation (kWh/m$^2$ Year) | Average Wind Power Density (W/m$^2$) | Geothermal Potential Conductivity (W/mk) | River |
|---|---|---|---|---|
| Bassano del Grappa | 1334 | 37 | 1 | Brenta |
| Kadıköy | 1507.66 | 36.44 | NA | Marmara Sea |
| León | 1643.56 | 43.45 | 1–1.1 | Bernesga and Torio Rivers |
| Lublin | 1125.47 | 61.99 | 1 | Bystrzyca River |
| Trenčín | 1182.15 | 50.55 | 1 | Vah River |
| Vidin | 1450 | 50 | 1 | Danube |

### 3.2. MCDA Questionnaire Results: GIS Layers Selection and Prioritization

Macro-scale analysis was conducted, which requires GIS-based city context data on resource availability, urban macroform, land usage, energy and e-mobility structure, energy service availability, and social structure of the city.

The results from the MCDA questionnaire were used for GIS layer identification, reclassification, and prioritization purposes. Results are shown in Table 17.

**Table 17.** Selected GIS layers and prioritization values considered in the analysis of each city.

| Components | Bassano Del Grappa | Kadiköy | León | Lublin | Trenčín | Vidin |
|---|---|---|---|---|---|---|
| Urban macroform | 16% | 17% | 24% | 14% | 21% | 16% |
| Residential- mixed use areas | | 7% | 7% | 10% | 11% | |
| Commercial areas | | 1% | 3% | 12% | 11% | |
| Green Areas/parking lot | 16% | 10% | 6% | 6% | 10% | 16% |
| Public administration areas | | 13% | 7% | 12% | - | |
| Social-cultural-educational-sport areas | | 13% | 8% | - | 11% | |
| Solar potential | | 8% | 8% | - | - | 14% |
| Solar generation | 15% | - | 8% | - | - | - |
| Hydropower | 16% | - | 2% | 8% | 10% | - |
| Wind potential | 5% | - | - | - | - | - |
| Population density | 7% | 11% | 6% | - | 7% | 5% |
| Biomass | 10% | | 6% | 8% | 4% | 10% |
| Ground coupling | - | 12% | 7% | - | 5% | - |
| E-mobility | - | 8% | 8% | 12% | - | - |
| Public domain | 15% | - | - | - | - | 16% |
| Heat grid | - | - | - | 9% | 6% | - |
| Hydrothermal potential | - | - | - | - | - | 10% |
| Geothermal potential | - | - | | | 4% | 14% |
| Waste heat potential | | | | 9% | | |

These city data layers play a decisive role in the creation of potentials and resources for PED implementation.

### 3.3. Weighted Overlay Result: Cities Candidate Areas to Become Positive

The last step in the identification of the candidate areas to become positive, consisted of an overlay analysis of the GIS layers included in Table 17 and supported in GIS software. The influence values that are presented in Table 17 are also supported by the reclassification and remap values that are obtained from the outcome of the MCDA Questionnaire. The collaborative work that is conducted with the municipality provided integration of qualitative and quantitative approaches for identifying PED Boundaries in six EU cities. The results of the most to least suitable areas are presented in Figure 3.

The exact size of the areas to be selected to become energy positive does not necessarily coincide with the official limits of the existing district.

Before the analysis, municipality representatives of each city preselected districts of interest to become positive. The criteria followed to make this selection was variable from city to city since the PED concept was not defined properly at that time. For example, Vidin selected two districts, Bononia and Himik, that have a concentration of both public buildings (mainly schools and kindergartens) and residential buildings, while Leon selected a degraded area that needs strong investment. Figure 4 provides details of Vidin's preselected districts by municipality overlayed with the results of candidate areas according to the Multicriteria and GIS analysis.

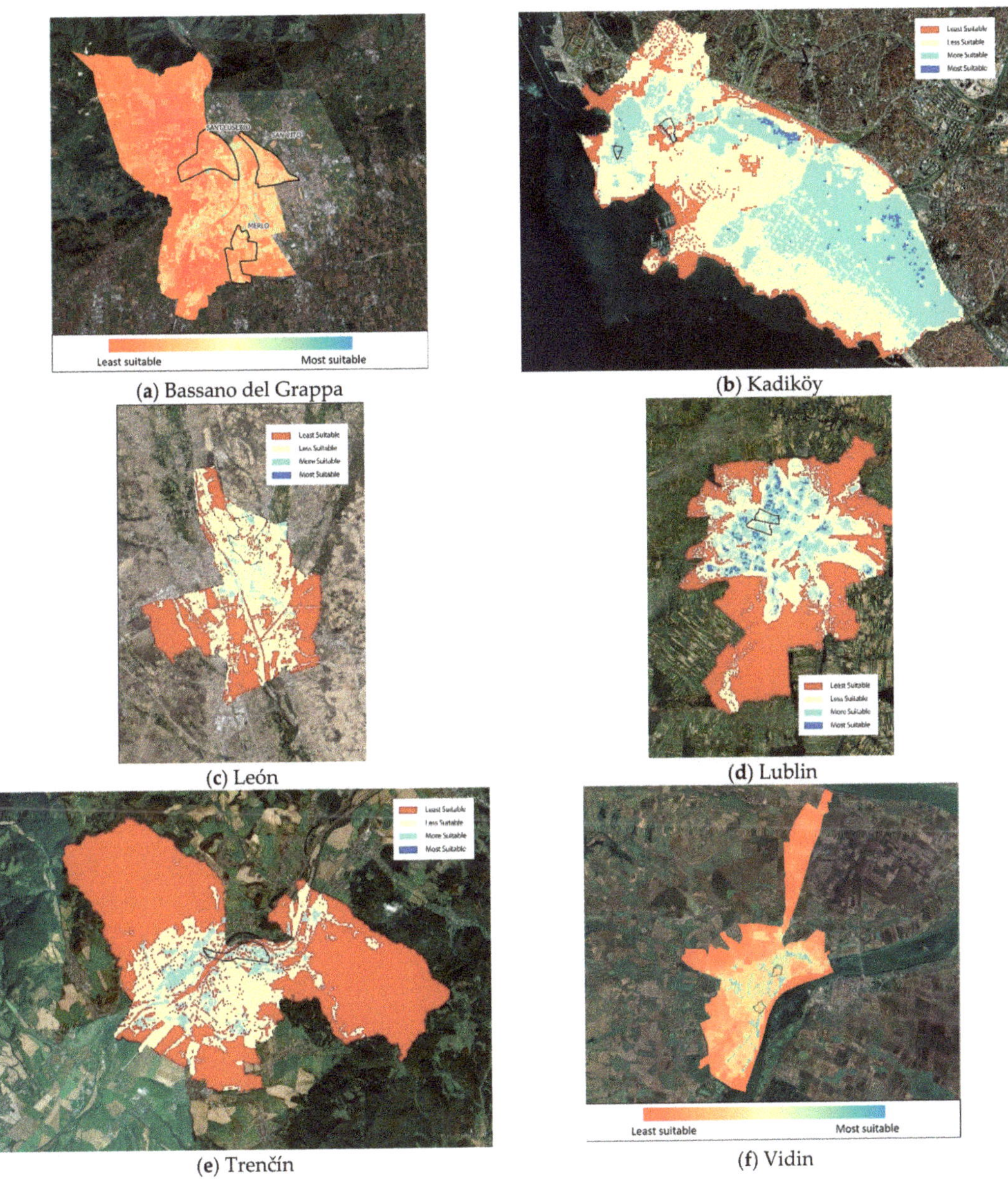



(**a**) Bassano del Grappa

(**b**) Kadiköy

(**c**) León

(**d**) Lublin

(**e**) Trenčín

(**f**) Vidin

**Figure 3.** Weighted overlay result for cities displaying suitable areas for PED implementations. (**a**) Bassano del Grappa most to least suitable areas; (**b**) Kadikoy most to least suitable areas; (**c**) León most to least suitable areas; (**d**) Lublin most to least suitable areas; (**e**) Trenčín most to least suitable areas; (**f**) Vidin most to least suitable areas.

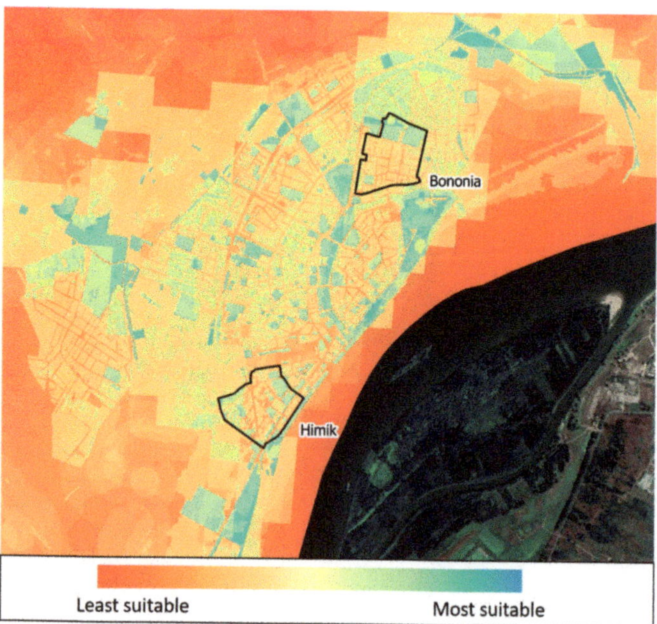



**Figure 4.** Vidin preselected and candidate areas to become positive.

All other results of weighted overlay analyses presented with the preselected PED areas are detailed in Appendix A.

## 4. Discussion and Future Developments

Based on the experience with the cities that participated in this research, the multi-criteria questionnaire ensures consideration of all the relevant criteria that can affect the correct implementation of a PED. Therefore, it allows the consideration of not only the renewable resource potential of a specific location, but also the economic opportunities, the legal framework in terms of enablers and barriers, the social context, and, of course, the existing and future modifications of the urban spatial form and planning, and the existence of or interest in creating infrastructure that can support the conversion to positive energy. Environmental issues are also considered, providing a deep overview of the whole city context. However, the criteria considered must be continuously updated in order to include the latest technological developments.

The proposed methodology develops a process for PED selection, which is also flexible enough to be adapted to each city's context in the sense that different choices can be made regarding each criterion. In the process of fulfilling the questionnaire in meetings with experts from the cities, it was well understood what was considered the "most suitable" situation and the status of the city in each of the criteria. Therefore, follower cities that participated in the process see themselves well-reflected in the results. On the other hand, it is recommended to fulfill the questionnaire with experts from different knowledge and experience backgrounds.

Regarding the overlay results, it must be noted that the maximum value obtained as a result of the overlay depends on the number of layers considered. In other words, introducing more spatial components makes it more difficult to reach the most suitable circumstances. This does not necessarily mean that the city has less areas with potential to become PED, but a more detailed analysis of the results will be needed to properly understand which criteria and/or combination of criteria affect the results and whether they are relevant for the assessment. For example, it could be more interesting to focus on several technologies and the renewable resources that support them than in trying to find a place with high potential in all resources.

It must be noted that in some cases, the connection between the multicriteria questionnaire and the GIS was challenging, and some criteria could not be included in the overlay analysis, resulting in the analysis being incomplete. The best example is the social criteria. In this sense, cities were not able to provide information about population projections per zone or even about the existing and/or potential energy communities. Therefore, including this information will enrich the assessment and the results will be more accurate.

With respect to the potentiality of GIS software, this research may also be further developed by the utilization of innovative tools regarding a diverse set of calculation optimization. Identified boundaries may also be enriched through the generation of digital twins by integrating BIM-based building models into the GIS-identified area.

## 5. Conclusions

The methodology developed for macro-scale analysis in cities to identify zones with the potential to become energy positive combines both the potential of multi-criteria analysis and the overlay analysis of GIS software. This combination provides a robust assessment and very visual results that can be easily understood at a quick glance. The process is conducted by active participation of the city representatives, both at data collection and evaluation phases, thus it is very participative and co-creative.

Results presented in this article provide a prioritization of the areas from the six European representative cities with the highest potential to become energy positive. The MCDA questionnaire helped the municipalities to identify critical criteria that affects both the cities (set city background) and the areas with potential to become positive (drivers and barriers to the implementation). The involvement of municipality representatives in several iterations of the process was indispensable to ensure correct implementation and for understanding of the results. This was considered critical considering that the municipalities made a preselection of districts before the assessment. The visualization of the results in GIS format was very positively received and allowed for accurate comprehension of the obtained results.

Results given in this article will guide the definition of PED boundaries in which micro-analysis will be the focus. Thanks to the PED Analytical components analysis, the first approach to understand the solutions suitable for the conversion of selected areas into a PED is also examined. Following this paper, further research will be conducted for micro-scale analyses and the outcomes will pursue a technology selection process.

**Author Contributions:** Conceptualization, B.A. and A.L.R.; methodology, B.A., A.L.R., P.H. and O.T.; software, O.T. and N.H.M.; validation, B.A.; formal analysis B.A., A.L.R. and P.H.; investigation, B.A. and A.L.R.; resources, B.A., A.L.R., and the cities Bassano del Grappa (Italy), Kadikoy (Turkey), León (Spain), Lublin (Poland), Trenčín (Slovakia), and Vidin (Bulgaria); data curation, O.T. and N.H.M.; writing—original draft preparation, B.A. and A.L.R.; writing—review and editing, B.A. and A.L.R.; visualization, O.T. and N.H.M.; supervision, B.A.; project administration, B.A.; funding acquisition, P.H. All authors have read and agreed to the published version of the manuscript.

**Funding:** This research was funded by the EU H2020 Programme under grant agreement n°824418.

**Institutional Review Board Statement:** Not applicable.

**Informed Consent Statement:** Not applicable.

**Data Availability Statement:** Not applicable.

**Acknowledgments:** Authors acknowledge the support provided by the city representatives of Bassano del Grappa (Italy), Kadikoy (Turkey), León (Spain), Lublin (Poland), Trenčín (Slovakia), and Vidin (Bulgaria) for sharing city GIS data for the analyses. The authors also acknowledge the funding received from the European Union's Horizon 2020 research and innovation programme under the MAKING-CITY project agreement No 824418.

**Conflicts of Interest:** The authors declare no conflict of interest.

## Appendix A. Candidate Areas to Become Positive

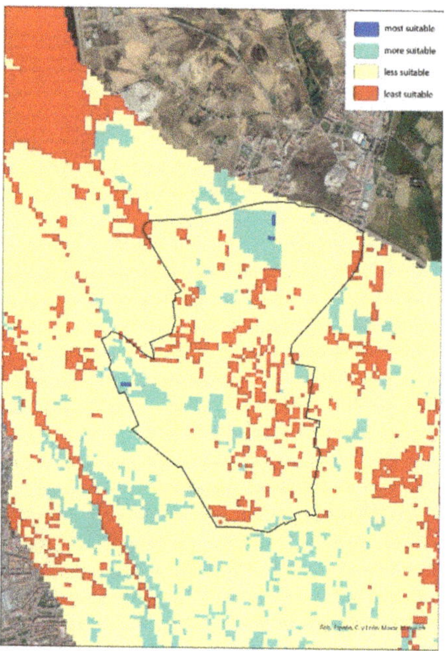


**Figure A1.** Leon's preselected and candidate areas to become positive.

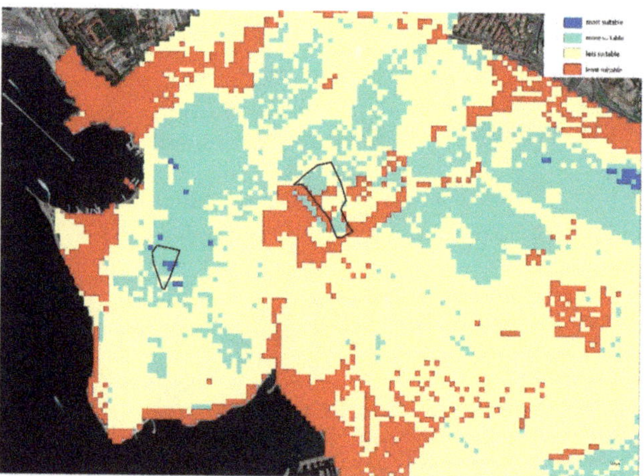


**Figure A2.** Kadikoy's preselected and candidate areas to become positive.

**Figure A3.** Trencin's preselected and candidate areas to become positive.

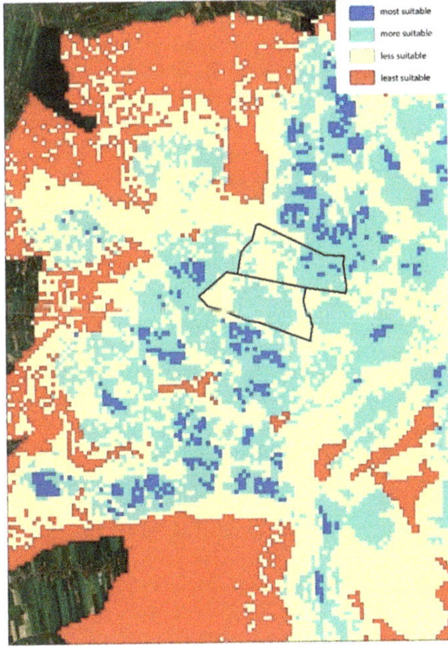



**Figure A4.** Lublin's preselected and candidate areas to become positive.

## References


1. Directive 2010/31/EU of the European Parliament and of the Council of 19 May 2010 on the Energy Performance of Buildings. Available online: https://eur-lex.europa.eu/legal-content/EN/TXT/HTML/?uri=CELEX:32010L0031&from=EN (accessed on 28 June 2021).
2. SET-Plan Temporary Working Group 3.2. SET-Plan ACTION No.3.2 Implementation Plan: Europe to Become a Global Role Model in Integrated, Innovative Solutions for the Planning, Deployment, and Replication of Positive Energy Districts (June 2018). Available online: https://setis.ec.europa.eu/system/files/setplan_smartcities_implementationplan.pdf (accessed on 28 June 2021).
3. Vandevyvere, H.; Ahlers, D.; Alpagut, B.; Cerna, V.; Cimini, V.; Haxhija, S.; Struiving, E. *Positive Energy Districts Solution Booklet*; European Union: Maastricht, The Netherlands, 2020.

4.  MAKING-CITY (The MAKING-CITY Project Received Funding from the H2020 Programme under Grant Agreement No 824418). Available online: https://makingcity.eu/ (accessed on 31 July 2021).
5.  POCITYF (This Project has Received Funding from the European Union's Horizon 2020 Research and Innovation Programme under Grant Agreement No 864400). Available online: https://pocityf.eu/ (accessed on 31 July 2021).
6.  ATELIER (This Project has Received Funding from the European Union's Horizon 2020 Research and Innovation Programme under Grant Agreement No 864374). Available online: https://smartcity-atelier.eu/ (accessed on 31 July 2021).
7.  +CityxChange (This Project has Received Funding from the European Union's Horizon 2020 Research and Innovation Programme under Grant Agreement No 824260). Available online: https://cityxchange.eu/ (accessed on 31 July 2021).
8.  SPARCs (This Project has Received Funding from the European Union's Horizon 2020 Research and Innovation Programme under Grant Agreement No 864242). Available online: https://www.sparcs.info/ (accessed on 31 July 2021).
9.  RESPONSE (This Project has Received Funding from the European Union's Horizon 2020 Research and Innovation Programme under Grant Agreement No 957751). Available online: https://h2020response.eu/ (accessed on 31 July 2021).
10. Girardin, L. *A GIS-Based Methodology for the Evaluation of Integrated Energy Systems in Urban Area*; EPFL: Lausanme, Switzerland, 2012; p. 218. [CrossRef]
11. Höhn, J.; Lehtonen, E.; Rasi, S.; Rintala, J. A Geographical Information System (GIS) based methodology for determination of potential biomasses and sites for biogas plants in southern Finland. *Appl. Energy* **2014**, *113*, 1–10. [CrossRef]
12. Díaz-Cuevas, P. GIS-based methodology for evaluating the wind-energy potential of territories: A case study from Andalusia (Spain). *Energies* **2018**, *11*, 2789. [CrossRef]
13. Mentis, D.; Welsch, M.; Nerini, F.F.; Broad, O.; Howells, M.; Bazilian, M.; Rogner, H. A GIS-based approach for electrification planning—A case study on Nigeria. *Energy Sustain. Dev.* **2015**, *29*, 142–150. [CrossRef]
14. Gaspari, J.; De Giglio, M.; Antonini, E.; Vodola, V. A GIS-Based Methodology for Speedy Energy Efficiency Mapping: A Case Study in Bologna. *Energies* **2020**, *13*, 2230. [CrossRef]
15. Lucchi, E.; D'Alonzo, V.; Exner, D.; Zambelli, P.; Garegnani, G. A density-based spatial cluster analysis supporting the Building Stock Analysis in Historical Towns. In Proceedings of the Building Simulation, Rome, Italy, 2–4 September 2019.
16. Gabaldón Moreno, A.; Vélez, F.; Alpagut, B.; Hernández, P.; Sanz Montalvillo, C. How to Achieve Positive Energy Districts for Sustainable Cities: A Proposed Calculation Methodology. *Sustainability* **2021**, *13*, 710. [CrossRef]
17. Alpagut, B.; Akyürek, Ö.; Mitre, E.M. Positive energy districts methodology and its replication potential. *Multidiscip. Digit. Publ. Inst. Proc.* **2019**, *20*, 8. [CrossRef]
18. Ryan, S.; Nimick, E. Multi-Criteria Decision Analysis and GIS. 2019. Available online: https://storymaps.arcgis.com/stories/b60b7399f6944bca86d1be6616c178cf (accessed on 15 June 2021).
19. Akbari, V.; Rajabi, M.A.; Chavoshi, S.H.; Shams, R. Landfill site selection by combining GIS and fuzzy multi criteria decision analysis, case study: Bandar Abbas, Iran. *World Appl. Sci. J.* **2008**, *3*, 39–47.
20. Kritikos, T.R.; Davies, T.R. GIS-based multi-criteria decision analysis for landslide susceptibility mapping at northern Evia, Greece. *Z. der Dtsch. Ges. für Geowiss.* **2011**, *4*, 421–434. [CrossRef]
21. Çetinkaya, C.; Kabak, M.; Erbaş, M.; Özceylan, E. Evaluation of ecotourism sites: A GIS-based multi-criteria decision analysis. *Kybernetes* **2018**, *47*, 1664–1686. [CrossRef]
22. Alpagut, B.; Montalvillo, C.; Vélez Jaramillo, F.; Hernandez Iñarra, P.; Brouwer, J.; Hirvonen-Kantola, S.; Kosunen, H.; Hermoso Martinez, N.; Ahokangas, P.; Rinne, S.; et al. MAKING-CITY D4.1—Methodology and Guidelines for PED Design; MAKING-CITY Project Report, Grant Agreement No 824418, Submitted to the European Commission; 2019. Available online: https://makingcity.eu/results/#1551708358627-aefa76ef-66b2 (accessed on 17 September 2021).
23. What Is An Inclusive City? Available online: http://www.inclusiveurbanism.org/ (accessed on 28 January 2021).
24. Understanding Overlay Analyses. Available online: https://pro.arcgis.com/en/pro-app/latest/tool-reference/spatial-analyst/understanding-overlay-analysis.htm#:~{}:text=Overlay%20analysis%20is%20a%20group,locations%20for%20a%20specific%20phenomenon (accessed on 28 January 2021).
25. Kahraman, S.; Unsal, O. *ArcGIS Spatial Analyst Handbook*; Esri: Redlands, CA, USA, 2014.
26. Hermoso, N.; Hernández, P.; Fernández, N.; López, A.; Alpagut, B.; Tabanoğlu, O.; Akgul, G. *Making-City D4.3—Analysis of FWC Candidate Areas to Become a PED*; Making City Project Report, Grant Agreement No. 824418. European Commission Submitted to the European Commission. 2020. under review. Available online: https://makingcity.eu/results/#1551708358627-aefa76ef-66b2 (accessed on 17 September 2021).

MDPI
St. Alban-Anlage 66
4052 Basel
Switzerland
Tel. +41 61 683 77 34
Fax +41 61 302 89 18
www.mdpi.com

*Energies* Editorial Office
E-mail: energies@mdpi.com
www.mdpi.com/journal/energies